Tor Browser

Learn How to Claim Your Privacy in the Internet World and Hide Your Ip

(Hide Your Ip Address and Ensure Internet Privacy)

Joseph Robson

Published By **Bengion Cosalas**

Joseph Robson

Tor Browser: Learn How to Claim Your Privacy in the Internet World and Hide Your Ip (Hide Your Ip Address and Ensure Internet Privacy)

ISBN   978-1-77485-472-3

No part of this guidebook shall be reproduced in any form without permission in writing from the publisher except in the case of brief quotations embodied in critical articles or reviews.

Legal & Disclaimer

The information contained in this ebook is not designed to replace or take the place of any form of medicine or professional medical advice. The information in this ebook has been provided for educational & entertainment purposes only.

The information contained in this book has been compiled from sources deemed reliable, and it is accurate to the best of the Author's knowledge; however, the Author cannot guarantee its accuracy and validity and cannot be held liable for any errors or omissions. Changes are periodically made to this book. You must consult your doctor or get professional medical advice before using any of the suggested remedies, techniques, or information in this book.

TABLE OF CONTENTS

Introduction

What exactly is Tor? Tor is a web browser which provides anonymity by concealing your identity while surfing the internet. It allows you to browse the internet as well as share content, and communicate with other users online while remaining completely anonymous. Tor is an abbreviation to mean The Onion Router and was developed in the US in the mid-nineties. Tor can encrypt all information sent by your computer to ensure that no one will be able to determine where you're from or even who you are. Tor gets its name 'Onion' due to the fact that encryption is built up of layers. The data you send from your computer goes through a number of'relays' or 'nodes' (other peoples' computers) operated by millions of people across the world, creating security layers. This is similar to creating the layers of an onion. Tor will conceal your IP address and provide you a new IP address each when you transmit the data. It's nearly impossible for anyone to determine which data came from.

The most straightforward way to access Tor is to use its browser (you can download it at www.torproject.org) that works for Windows, MacOS and Linux (choose the appropriate download). The Tor browser was developed and built in part on Firefox browser, but it blocks some of the plugins that could affect privacy and security while browsing the internet. Your firewall and antivirus software might need to be reconfigured to allow you to connect to Tor. Tor network. It is also possible to use the Tor application for your Android phone. The app is known as Orbit and it also comes with an operating system known as Tails which is configured for Tor.

Tor is a well-known browser that is used by police as well as military and government organizations all over the world. There are human rights campaigners, medical researchers journalists, whistleblowers, journalists, and even terrorists who use the Tor browser. They all share the same goal which is to guard their privacy, their communications, and data from the prying eyes of others. When you use Tor you are able to choose which you connect with. It is a huge community of Tor users. Facebook's only Tor version of the site is becoming highly popular with over one million users per month.

Tor is a legal program and was not designed to be used for illegal activities. The users of Tor that could abuse the power of Tor to engage in criminal activities and trade in illegal substances. It is the same of any other web-based service Users are the ones who commit the acts. There are more legitimate users than are criminals and there's nothing wrong with protecting your privacy. There is no need to worry about security since Tor doesn't offer an index of sites that are considered to be dark. You will not come across any harmful or illegal content unless it's a recognized web address using an .onion domain.

Tor has been designed to safeguard the privacy of the user by allowing them to use private communications and stay clear of monitoring by the analysis of traffic and network surveillance. It was not created with criminal activities in mind.

4

Chapter 1: Protocols

In the earlier chapters there were references to protocols as the method by computer systems communicating with one another. Human beings, when conversing in a conversation, don't need to be rigid with the usage of the language. We all can discern how the person speaking is when they just communicate their thoughts. However, this is not the case for computers. The method by which computers communicate is controlled electronically. The precise mechanism by which this happens is a bit nebulous.

The first step in the computer chat is "handshaking. This creates connections between the computers. Simply put this signifies that computers are establishing an internet connection. The IP addresses of the computers are synchronizing , so they can achieve this. What happens next is determined by protocols that are applicable to different kinds of Internet actions. In this instance, it could be as SMS, email or web browsing. The transmission of information via the Internet typically requires the message to be broken down into smaller pieces, which are reassembled after they reach their destination. This makes it simpler to transfer the message over to another side. If the message is not split into several packets, the message may take years to get to the computer it's connected to.

TCP The protocol that divides the data into packets to be transmitted. It allows them to be transmitted. When they are delivered, they break into small packets. When it reaches the other side, after being handed over to the client it reassembles them to ensure that there are no lost packets in the transmission. TCP is a shorthand in for Transmission Control Protocol. The acronym is fairly simple to understand.

IP: IP stands for Internet Protocol, which directs information to the right address. Each computer that connects to Internet is assigned a unique unique address. This is vital when we begin discussions on the safety of online browsing. IP is not able to have a connection to computer. That's one of the main functions of TCP. It manages only the channels through which packets travel. Any activity that occurs through that IP address is viewed by other people if they look deep enough. They'll be able discover what you did after you have looked at it online, and also the way you conducted your business. You can erase your hard drive from your PC, however, your IP address is going to remain to keep that information within your PC.

Pop3/IMAP: One the most valued components that make up the Internet is email. Email is controlled by protocols. The protocol used to send emails is SMTP that stands for Simple Mail Transfer Protocol. The protocol for receiving emails is either POP3 that stands for Post Office Protocol 3 or IMAP that stands in for Interactive Mail Access Protocol. You might think it's simple as a click of a button. However, there are numerous protocols involved in sending and receiving an email.

HTTPor HTTPS HTTP/HTTPS: World Wide Web is constructed of files using the suffix HTML. Web pages are controlled through the protocols HTTP that stands for Hypertext Transfer Protocol. Copy and paste the URL from the address bar in your browser, you'll likely encounter a sentence that begins with HTTP ://... The HTTPS part is a bit more common. In some instances, more then others. You may be able to see HTTPS as opposed to HTTP. HTTPS stands for Hypertext Transfer Protocol Over Secure Socket Layer. It means that any data sent on this site is encrypted form and theoretically inaccessible. We'll have more to discuss this in the future.

FTP: This abbreviation refers to File Transfer Protocol. It is not likely to be an issue in the event that you upload files to websites by using FTP software such as FileZilla.

This concludes our brief overview of the variety of protocols that you may encounter while using the Internet. There's more than could be written about the subject. In fact, it will take more than just this chapter to cover each one. What I've presented are the primary elements that will aid the Internet accomplish what it is required to accomplish.

Chapter 2: Are You Being Tracked Online?

Are your online activities being observed? It could be that you don't know? It's common for people to click something on a website that they are' interested in: an article of clothing, a book of clothing, a music download and so on. After that and you are able to see ads appearing on every place like Facebook or mail, Google etc.

You might find this to be extremely annoying and should install an ad blocker that could assist, but it could also cause problems for you. The reason clicking on an item led to the rush of ads was because your internet activity was being tracked , even although you were unaware of that it was happening. There are those who have set up blocks, so whenever a specific item is clicked, they'll be aware. All their processed data like web pages , will be able inflicting damage on your PC.

One of the factors that have been anticipated is the increase of individual prices. If you pay much for an product could be taken as an indication that you'll be paying a high price for other items. It is possible to purchase music, and the cost you pay will be more expensive than that of the other person, whom you believe to have less.

Brick and mortar businesses are also collecting data about the people you. If you've got any loyalty cards, each time you purchase something, you will discover that information regarding the purchase being recorded in secret. It is possible to receive emails or other messages that contain information on other products that might interest you.

Another fact that is often not known is that shops can track Wi-Fi signals from customers and can determine what aisles people can explore. This data is mostly used in the shop to decide the layout of the products. If you don't wish to assist the shop in finding the place you're going and where you are, turn off your mobile before you step into the shop.

The data that is being gathered may include information you've entered yourself , like your name address, address, and details of your credit card. Additional information can also be gathered, such as information about your device, websites you visit or your online activities, among other information.

There are different ways that tracking is being used. A giant in defense called Raytheon has come up with a piece of software known as"RIOT. It is a targeted attack on Facebook and makes a profile of an individual who is interesting to. It is focused on such things as logins and longitudes, latitudes, additional information taken from images and many other factors that allow an idea to be built to be built.

In the past, this monitoring was very difficult and therefore only targets of particular interest were monitored. With the advent of big data and artificial intelligence that allows patterns to be identified in the massive amounts of data, it makes sense to conduct mass surveillance.

The concept of this study is, by identifying the behavior patterns of normal people, the behaviour of the people of concern: the child abusers, criminals terrorists, child abusers, etc. will be revealed. The argument is that if you do not have anything to hide, you've got no reason to be concerned and this could be a reasonable assumption.

But, if the information that is trawled by software like RIOT ends up in the wrong hands, then serious issues could be triggered for innocent people.

Another technology is being employed by those who wish to track you on the internet. It's completely different from the concept of RIOT. It is referred to as canvas fingerprinting. It's based on an extremely clever concept that each of the computers have fingerprints, and these are distinct.

The canvas fingerprinting software allows your computer to transmit a photo of text. Every computer's text is unique to the computer. This is actually a monitoring of the device, not the individual. But, if just one person is using the device, it's actually an examination of the individual.

Websites in size from YouPorn up to White House have used this technology. YouPorn claims that they no longer make use of the technology. Both of the technology and other technologies suggest that you'll always be tracked whenever you go online. There is a belief that at most 5percent from the most popular 100,000 sites employ canvas fingerprinting, which includes US as well as Canadian government websites.

Every organization does the following of your online activity: social media mobile phone companies, email services, mobile phone apps, search engines, and so on. These organizations are keeping track of your online activities.

The compilation might appear harmless, however if the information is in the wrong hands, such as government spy agencies mafia, the mafia, or another criminal organization, or even a spy, it could have catastrophic consequences like unlawful arrest and identity theft, online ransom , etc.

There are many ways to keep the release of this information. The purpose of this book to explore these. Different methods for protecting your privacy online will be reviewed with special focus on one approach that involves making use of the TOR browser. This browser will be discussed extensively in Chapter 4. The next chapter we'll look at privacy online as a whole.

Chapter 3: What's The Best Way To Stay Anonymous Online

IP Address: Before it was mentioned that the unique address your computer owns, also known as IP Address. It's an individual identifier of your computer. There are two different versions of this: it's the IPv4 in addition to the IPv6. There were a number of identification numbers feasible with IPv4 ended and IPv6 was created. IPv6 became available in the year 2015. You don't need to be aware of the specifics of these numbers unless you're curious. Make sure to remember that these numbers are available and very useful to individuals and companies that want to monitor your activities while you're online.

Mac Address: A different set of numbers linked to your computer, and must be concealed or concealed are your Mac address. It is the number that identifies the network adapter that is located on your computer. It is made up of a series of numbers that are which are followed by colons. If you're interested, you can look up the methods that will reveal you Mac address.

In this sense, Mac does not refer to the Macintosh computer. It is an address that is the Media Access Control address of your computer. All computers, no matter if they are running Windows, Macintosh, iOS, Android or Linux operating systems will have these addresses when they are connected to Internet that is what most computers today do. Much like the IP address the Mac address is of particular significance to those who want to track your online activity.

The previous chapter outlined ways to stop others from spying on your online activities. If a major company with huge resources, such as the CIA would like to follow your activities, they will and you must be extremely careful to stay out of this. However there are some suggestions to remain as secretive as you can be:

1. Utilize private sessions whenever you browse the Internet. The majority of browsers like Firefox, Chrome, and Safari are able to do this. Be aware that private

browsing is only for use in private. In any work or school environment the administrator of your network will be able to see what you're up browsing on the Internet.

2. Logins: Use different passwords for different sites. It's a hassle, but it's better to keep a variety of passwords and perhaps an account book to store the passwords in, than having your credit card stolen by a hacker to make the purchase of $10,000 in Manila!

3. Clean out all cookies and erase frequently. Cookies are small files that are placed by many websites in your browser to serve a variety of purposes however, the majority of them are harmless. This is, however, exclusively for use in domestic settings. Don't believe that just since you've erased cookies, you're safe online.

4. Don't let your browser send location data. If you don't want to have your location disclosed, this is an easy choice. Every major browser has the capability of preventing them from transmitting your location information forward.

5. Don't let Google track you. Google is an excellent search engine that is utilized by the majority of Internet users. But, it's best to stop Google from monitoring your activities. If you don't take this step you'll be in the hands of advertisers. In the future, we'll inform you about the Epic browser, which blocks Google's tricks.

6. Make sure that you set your social media accounts such as Facebook, Twitter and Linked In privacy settings so that you can enjoy the most privacy. It's a bit scary to realize the volume of personal information accessible on social media sites like Facebook and Twitter and this is the cost that you pay to access an absolutely free service. Try using the Facebook settings, and download copies of your Facebook information and you'll be able to see every single activity you've ever made on Facebook that you have ever done in your personal records of data. Similar potential for data harvesting across all social media platforms so the best way to stop this happening is to remove all accounts. Deactivating your account will not erase the

data, but it will to put it in hibernation in case you want to reinstate your account.

7. Make use of an add-on, such as Privacy Badger to stop trackers. Privacy Badger is a browser-based add-on tool that can identify websites that might be trying to monitor your activities in an unconvincing and indecent way. The copy you download from Private Badger keeps track of any third-party domains which are in use on the sites you visit. They may track users without your consent using cookies to create the details of your internet actions. Privacy Bandit will not allow any content from any third-party tracker except when the domain is significant roles in the structure of the site for example, images or maps that are embedded. In these cases, Privacy Badger will let connections but remove the potentially harmful tracking cookies.

There are other add-ons available, such as AdBlock Plus, AdBlock ultimate and Tinfoil for Facebook It's worth it to look for the appropriate software that is compatible with your preferred style of browsing.

8. Remove Java, JavaScript and all plug-ins you don't utilize. Because JavaScript is so popularly used on websites and other websites, it's not easy to accomplish in all situations. If you visit travel websites, which usually employ a large amount of JavaScript be cautious.

9. Make use of to use Epic browser. It is a variant of Chromium which is quite like Chrome that includes many of the features that have been endorsed prior to inclusion into the web browser. It also shows the number of trackers who tried to track you every day. The browser is rapidly growing into one of the top well-known internet browsers. It is also incredibly popular with people in China and the US as well as Nigeria.

If you use Epic data is encrypted automatically which means that activities aren't monitored. It is an excellent tool for anyone who wants to gain access to information that might not be able to be authorised by, for example local government. Furthermore, when making use of this Epic browser, the irritating advertisements that pop up when you search online are blocked, which means you don't have to spend always trying to clear the ads off of your display. The only drawback to using Epic is its speed. Typically, when you are searching for pages you're searching for with Epic you might get a slower result , and the pages could be slower to download, which is a tiny price to pay for the extra security. Since this is a no-cost download, it's an excellent option to include in your arsenal of online security.

Epic has been fighting an intense struggle with Google which is dependent in their capability to make ads sell for large portions of their income. Google has been trying its best to dismantle Epic. Their conduct in this case is very similar to other IT giants, when something threatens the gold mining operations of their company.

Since Google offers almost unlimited resources and resources, this is definitely an David against Goliath battle and, in addition to installing the browser, I'd like to recommend that you follow all of the other suggestions I previously mentioned so to be able to browse in complete anonymity.

Apart from its outstanding work to stop your identity from being traced through Google, Epic has a built-in method of blocking canvas fingerprints, a method of tracking that was discussed in the final chapter.

10. Utilize use a VPN. This is a reference to Virtual Private Network. If you truly want to be secure make sure you spend anywhere from $5 to $10 per month for a reliable VPN.

A VPN is an encrypted server that performs the tasks through the Internet that you request it to perform in such an approach where your address remains concealed. One of the best advantages that comes with VPNs is that, when you access an internet page with the HTTP protocol and then enter the password or engage in other secret procedures, the Internet traffic that is generated isn't secured. If you utilize the VPN and you use it, then every actions you can perform in your browser is immediately secured. It's as if that the connection was http:http:///

VPNs are subject to the most scrutiny in the eyes of US authorities. If the VPN server is located in the USA is much more susceptible to being vulnerable than one that is located in other countries like Sweden and Germany. But however, if your desire to be anonymous is so that you could watch children's porn or commit terrorist acts, you'll eventually be arrested. The way in that online activity is monitored is always evolving.

11. Spoofer for Mac: Spoofer: MAC (Media access control) addresses are unique to every device and are easily identified. To protect yourself from this, you can utilize software known as the Mac Spoofer. There are numerous reasons to alter or fake the MAC address, with the most obvious reason being to bypass the restrictions of your network and provide you with an additional level of security.

Another advantage of the fake MAC address is it is an alternative to an unreliable router. By using the fake MAC you will still be able to be able to access the internet. If you want to alter the MAC address, launch your start menu, then select your control panel. After you have launched your control panel, select the internet and network option. Click on network and sharing center, usually the first option available and then click on it, this will lead you to your communication/network setup and connections, select change adapter settings.

Choose local area connections, click on properties . Then click on the Configure option in the Configure window, look at the top of the screen where you will see the Advanced option. Click on it. Under the Advanced options, you will find an additional window titled Settings. Then scroll to the bottom until you reach Locally Administered Address, click on it. Find the text that will appear in yellow background. Search for the text that should closely match the MAC address of this network adapter. Just next to the Settings window is a field labeled Value. The user will type an entirely new set of characters to deceive your MAC address.

It is important to note that prior to changing the MAC address, it is beneficial for you to examine the match between your old address. Now, go into the menu called Start. click it. There is a search bar located at the bottom of the screen, just below All Programs. In the search bar, you can type cmd. Then, scroll towards on the right side of results, and you'll see something marked cmd.exe Click on it. The new screen with black background will pop up with a few lines of text. There is also an underscore with a blank which will flash in and out. This is where you type in getmac , and hit enter. The result will be the list of physical Addresses being displayed one of which is the currently MAC address. Open the Advanced options window . In doing so , you can modify you MAC address and create a fresh one by following the format of your previous address. It will contain 12 characters total.

It is possible to use combinations of letters from A-F , as well as any other numbers and under Value, you can enter the new address, be sure to preserve the format. For example for example, if the initial 4 characters are A3-E2, then you can alter the characters to E1-D2. By repeating the process for all 12 characters. make sure you adhere the formatting that is required. Now, the last step is to click OK at in the middle of the screen The Advanced option window disappears from your screen and you will see movement within Local Area Connection. Local Area Connection and for some time, the red cross will be displayed on the left side of Local Area Connections and will show the word Disabled. This is simply a sign that the system is recording an update and after couple of seconds, this will disappear and the LAPs are going to become activated. Congratulations! You've successfully faked your MAC address.

12. Take advantage of the TOR browser and you'll not be dissatisfied. This is the principal subject of the book, so I'll not go into detail about it at this point. The next chapter is filled with information on it, and the best way to set it up, and make use of it. It will be a pleasure for it, as it will allow you to be able to use the internet and feel safe using it.

Additionally, you should ensure that your anti-virus and anti-malware software is current. In this article, I'll define the difference between these two programs and the steps you need to follow to select the right product for your PC and you.

What exactly is malware and what is malware? It's all kinds of viruses that could affect your device. Spyware is software designed to gather data without your permission and then transmit the information to third parties. Adware is a different type of malware that's specifically designed to create advertisements so that it can generate income for the creator. The Trojan like its title, is a destructive program that can use your software to deceive users about its real intention and is usually spread through social interactions, or perhaps an e-mail attachment which contains the standard form.

A virus On the other hand is a bit of code that could copy itself and harm your device. So, although all viruses are considered malware but not all malware is malware.

The primary distinction between anti-malware software and antivirus is that an antivirus program will be most likely to focus on old, more well-known threats like Trojans and viruses whereas anti-malware software is designed to tackle more recent and modern threats. While your antivirus software is working to fight malware, you could be exposed to traditional sources like email or even a USB.

Which one should you choose to use? The most straightforward answer is to run both applications side-by-side since when you attempt to mix the two components, you will lose some aspects and since anti-malware programs are generally light and simple to use, it's designed to be used in conjunction with antivirus programs to provide security layers against malware and viruses.

The most effective health plan for your PC could include an anti-virus program like Bitdefender, Norton by Symantec or Kaspersky Anti-virus which are the top three rated the year 2017 in PCMag. To find a reliable anti-malware software, look into

Hitman Pro, Malwarebytes, Zemena or Emsisoft. Another tip when selecting your program, do not trust untested malware/adware or viruses removal tools since they could also cause damage to your computer and create the type of issues you want to avoid.

Chapter 4: The Tor Browser

TOR is the abbreviation used for The Onion Router. It was initially an international network of services created in the US military to ensure it could ensure that Internet browsing and usage could be conducted without revealing the identity of the user. TOR transmits Internet traffic through relays an international network that includes thousands of servers to conceal the location and activities of a user from monitoring and analysis of traffic.

The TOR servers, which serve as relays, are operated by volunteers who are keen to safeguard the privacy of online users and ensure security. Instead of a direct link to the origin or destination of their data the users' network communication pass through a vast amount of different servers. This can cause to make it difficult for trackers to identify the source of their traffic.

Anyone who would like to protect their Internet data out of reach of journalists, advertisers and other individuals, will gain from the use of Tor. It's an excellent tool for undercover police officers working in criminal groups. It allows them to go where police officers are totally unnoticed as they conduct their investigations.

It is a great tool to combat the censorship. It lets users access sites that might normally be restricted. The privateness of the individual is protected. Some people misunderstand the desire to protect their privacy online as a sign of involvement in child porn or terrorism.

It's impossible to be further from the fact. Journalists living under a state of war would need anonymity online to ensure his or her freedom , and perhaps even life. They'd need anonymity to safeguard their sources.

The TOR protocol is extremely useful in the creation of a new way of communication that comes with privacy. On the normal web it is not possible to keep your information private. But with TOR you'll be able to surf the web anonymously and not have to worry about someone snooping on what you're doing.

TOR is home to what are known as hidden services that are referred to as hidden. It allows users to establish chat rooms. These chatrooms are where honest conversations can take place about topics like rape domestic violence, various types of ailments and whistleblowing. Information can be freely exchanged which can be extremely useful to insurance companies, large media, and business.

If one is employing TOR and someone attempts to track them down, then all the tracker will see is scattered points in the TOR network. The user's computer is not recognized. This will prevent your computer from being compromised which is a major issue in our modern society. It will allow you to surf the internet without having to worry about being targeted or not.

To use this network, it's required in order to download TOR browser and download it to your PC. The TOR system can result in more slow browsing than usual because of the number of relays through which transmit signals.

The services are available only to TOR users. Unfortunately they, though may have been created with the most noble of reasons, have also been extremely misused. There are numerous websites, such as The Silk Road whose purpose was selling illicit drugs, or the one that was discovered by the FBI which was the largest child porn ring which was discovered has set bad precedents. Despite all this, TOR has been endorsed by various organizations, such as Indymedia to safeguard their journalists as well as their journalists and the Electronic Frontier Foundation for upholding the privacy of online users.

Certain large companies use TOR commercially to monitor the behaviour of their competitors and to safeguard their own operations. The TOR protocol is superior to most VPNs since it's impossible to establish the time or the amount of communication.

It is still being used in The U.S. Navy and it is also used by law enforcement agencies for many of their operations, when it is essential not to change an IP address of the government.

The more people that use TOR is the greater and more secure it gets. Every user's network activity is hidden from the view of the other network users. A tracking device, that protects the user from, is known as Traffic Analysis. This method of sleuthing provides information of the origin, location, the time and duration of the communication. It also allows deductive reasoning about who has been communicating with who.

The method of analysis is to focus upon the head of the data packet. Data packets contain payloads, that could include emails video file, an email message or a PDF file. The payload can be encrypted, but the headers are not which is a security risk. A careful analysis of headers could reveal an abundance of data, including the destination, source and time, as well as the dimensions of payload.

The issue that people who desire privacy is the need for the computer who are receiving your message to access the data contained inside the header. The information could be accessed from other people using sophisticated software. This includes ISPs as well as law enforcement agencies, when they track you, as well as other trackers.

TOR minimizes the chance of detection by any kind of traffic analysis due to its distribution of messages via multiple servers, in order to keep trackers unaware of what the communication came from or to where it is heading.

The method by which this network of servers or nodes is created is extremely clever. It is constructed server by server and encrypted in the course of. Each relay and server does not know the entire route.

A circuit's completion allows various applications to work within the TOR network. TOR operates only on TCP communications, and includes SOCKS support. SOCKS is an Internet protocol that exchanges data between servers and clients via a proxy server. It is a server that serves as a connection between two networks.

The TOR app is fantastic, however it's foolish to think that by using it, you're totally private. If, for any reason, are brought to the attention of law enforcement the law enforcement agencies will eventually catch you. The only thing TOR could do is to slow this process.

One method by which organisations such as the FBI utilize to use the TOR network to track down bad guys is by joining TOR. Services of FBI servers are utilized as component of TOR network that is unnoticed by TOR.

However, the massive arrests of child pornographers, as previously mentioned, was not made through this method instead, it was through the injection with malware inside the browser of an victim who fallen victim to his own negligence. It is the browser's weakest link in relation to law enforcement and other monitoring. Attacks against the browser are known as "Man in the Middle" attacks.

Law enforcement agencies must be extremely cautious in how they conduct themselves since people who are using TOR represent the smartest individuals in the world. when it was discovered that law enforcement was taking too much of them in and taking countermeasures, that would result in TOR users safer and thus make them the law enforcement agencies more difficult to track down. Law enforcement agencies are therefore extremely cautious to hide its tracks, and only pursue the true criminals.

We will then examine ways to set up this browser. TOR browser.

The TOR browser is a variant of the free-source Firefox browser. Installation of the browser is simple. Google"download the TOR browser. It will take you to the download page of TOR. Follow the instructions. The menu will appear. of TOR downloads that are compatible with the different operating systems which are in fact Windows, Apple, Linux smartphones, and one with source code.

Choose the option you prefer. If you have selected Windows then you will be able to select the location you wish to put the browser into. Once you've done that, you can press install, and the installation process is done automatically.

There's one last window to be working on. Make sure to press the upper button until you are sure that you are sure that your Internet connection is blocked or filtered. We usually hit the upper button that states connect. The lower button is for configuration. Be sure to review the information on the TOR window.

If you press connectto the network, your connection will not instantaneous, but once it is finished, you should see a green screen with safety alerts and details about the TOR project. To make sure you're connecting with this Tor network, enter www.whatismyip.com in the address bar, and press enter.

It is evident that the speed of the browser compared to the normal speed that is Firefox and Chrome. Be attentive to the instructions TOR advises users to adhere to. One of these is that you should not Torrent when downloading files. If you do this when downloading an application, it can allow a virus, or hacker, a more easy gain access to the computer. Next chapter address what is known as the Deep Web which the TOR browser was specifically developed to protect against.

Chapter 5: Secrets Of The Dark Web

Dark web: Many people mistake with the Dark Web with the Deep Web. The dark web is the term used to describe the encrypted network that connects those servers at TOR and the users of the servers, namely the clients.

However, it is said that the Deep Web is composed of all files that exist that are on the Internet and cannot be indexable by standard search engines, such as Google as well as Bing. It is believed that the Deep Web has been estimated to be 99percent of all Internet content. According to estimates, it is comprised of 7500 Terabyte of information.

The majority of material found that is available on the Deep Web is not very fascinating. For instance, newspapers have Deep Web databases of stories that have not been published. Many other similarly boring data is also available. For example, the sales records of billions of transactions conducted by companies would be stored in Deep Web sites. This Deep Web is mainly benign and boring.

The Dark Web is far more fascinating. Anything you need is accessible via the Dark Web. You can purchase are:

1. Books banned.

Many people believe that they are on the Dark Web is filled with pornography drug traffickers and everything criminal, it's a little widely known truth that there exist an awe-inspiring amount of bibliophiles using the Dark Web. In 2011, Dread Pirate Roberts who founded Silk Road in the beginning Silk Road began his drug bazaar book club. He's quoted as saying "Knowledge is power and reading is among the most effective ways to increase your knowledge. Every week, we'll choose the book that will enhance our understanding of the challenges facing members of the Silk Road community and have an informal discussion on the material. I would like to see an open and lively discussion will be cultivated and that as a group we will become more steadfast in our beliefsand have an unifying message and voice, as the world starts to take note."

The world took notice, and Roberts was detained, in no way for Silk Road's bookclub, but rather for the trafficking of drugs and for money laundering. Silk Road was closed and the book club that it was a part of stopped anymore. However, they didn't have to be silenced. They were only one month later, Silk Road was up and operating again, and so was the book club that was being run by an Silk Road senior moderator by the name of Inigo until Inigo himself was detained and Silk Road decided that it would end its ties to Silk Road and continue in an anonymous chat room, while maintaining the tradition of Inigo who was, after all. The power of knowledge is in the mind! !

2. Credit cards copied from credit card stores.

The Dark Web is like a shopping center for cybercrime. It is not just about credit cards. the spammer lists, phishing kits and the entire tools required to commit all kinds of cybercrime are readily available. A large number of credit cards issued by US retailers are available on what's referred to as a carder's forum. it's like Craig's List for hackers where cards that are backed by customer information are traded for just $1.00 per card.

3. Fake passports.

There are numerous options to obtain fake passports via the Dark Web, the quality is, however, varying. A lot of people claim to have bought and used the fake passports at various times, and are supported by sellers who claim to have insiders from different departments which allow the fake information in legitimate ways. In the majority of cases the people who are involved are in a way that is illegal and require a reputable means of transportation, leading to them being able to locate those who sell illegal drugs.

4. Illicit drugs.

While only a small portion of purchases of illegal drugs are done online, the number is growing and it is increasing quickly, altering the entire process of dealing with drugs within the process. Sellers are now focused on offering a higher quality product at a lower price , thereby making a brand that is reputable. Estimated sales in 2012 were

$15m to $17m, but by 2015, this was up to $150m-$180m.

5. Hackers

Are you interested in finding out what your ex-partner is up doing? Maybe you want to take down a major business or bring it to stop, there's numerous websites in the Dark Web where you can locate people who will perform whatever you ask to ask them to do.

6. Burglars

It's not as common However, There are reports about websites in the Dark Web where you can hire someone to steal an order. The request is made with the most precise information you can. He will steal the item and provide you with an image to show that it was stolen. Evidently, he has the list of items up to sell that were purchased in the event that people placed an order for the thefts but never followed up with a payment.

7. Match fixing and illegal betting

Today, we have highly controlled betting practices, particularly due to the rise of online betting , which has grown exponentially since its beginning during the latter half of the 1990s and, according to Statista was responsible for a total which was worth $46 billion last year and was projected to increase to $56 billion by 2018. it's becoming increasingly difficult to make illegal bets. In the Dark Web however you can place all sorts of bets illegally as match-fixing is widespread. By clicking the mouse, you are able to practically do what you like.

8. Hitmen

This is an extremely dark. The possibility of hiring an experienced gun is common and while it's not inexpensive, it's not expensive enough to discourage serious buyers. One website claims to promote an assassination service for America or Canada for $10,000. US and Canada for $12,000 and $10,000 in Europe according reports in the Mail Online. Certain of the phrases and advertising techniques are frightening. One self-styled assassin boasts, "I do not know any information about you, and you don't know any information about me. The victim of choice will die in death. Nobody will ever know the reason for or who did this. In addition I do my best to make it appear as if it was a suicide or accident" It's a bit scary to think that there are people there who are in support of such a thing. But, you'll be shocked to discover that someone close to you may be involved in this kind of thing.

Most sales occur via "crypto marketplaces" the Black Web's similar to Amazon and E-bay. They use the same feedback process and allowing users to give ratings to sellers, items, and sellers and let other buyers make their purchases based on information they have gathered. Administrators get a cut of every sale, in order to make moderators pay (in bitcoin, of course) to manage forums and customer complaints.

9. Weapons.

There is speculation the fact that The Armory is the biggest and most well-known online market for guns, with the minimum purchase of $1050 It claims to have around 400 items to sell with a focus on firearms that are not traceable or possess a fake serial number. Additionally, they have an army section, they are believed to be among the biggest online sellers of weapons and have a huge following.

They must be paid for using Bitcoin which is an digital currency. Bitcoin is a peer-to- peer system developed in 2008 and utilized to pay for online transactions, without the help of a central authority that is trusted. Since its inception it has grown into something that goes beyond an ordinary currency. The bitcoin currency has its own user community and also serves as an investment tool. The key to bitcoin's success is the enormous peer-to- peer network and the consensus that allows for a system of payment that cannot reverse payments accounts can't be frozen, and that results in significantly lower transaction fees.

Bitcoin like the Internet it does not have a central administrator or owner. It is controlled by developers who invest their time into making sure that Bitcoin functions as it should. It is in their best interests to ensure that correct decisions are made. The influence that their contribution has is contingent upon the volume of computational power they give for the Bitcoin network.

Additionally, some users contribute their time to help in the smooth operation of the peer-to-peer network, and get rewarded with bitcoins that can be used on the internet. This is a basic type of mining, the word used to describe the process of getting bitcoins. The best method to learn about bitcoin is to purchase an amount and try it out There are a variety of methods to accomplish this and there are plenty of resources available to assist you.

Some people are foolish enough to have put sensitive information, like naked photos, on the Deep Web. They believe it's secure. But it's not!

One of the most famous examples can be found in an example of the Ashley Madison site, which was designed for spouses who were bored and wanted extramarital affair. Hackers hacked into the website and 10GB of information from the site was posted in the Dark Web and thus became accessible for users on TOR. It was discovered through a reporter named Brian Krebs who had written for years about security on the internet as well as the theft of data from major corporations through a well-known blog. He was investigating various companies such as Dominos Pizza, Tesco, and Adobe and was sent unidentified links to a cache of data taken from the Canadian company called Avid Life Media (ALM) of which he was only vaguely aware. Since 2008, they've operated an extremely well-known dating website that was specifically designed for married couples.

Offering absolute discretion at the time of the tip-off , they claimed at present to have 37.6 million users around the world. In the end, by simply clicking the hyperlinks he'd received a message Krebs and was able to look up actual credit card numbers for genuine members of the website which had previously been able to guarantee total discretion. Within the documents, he was able to find not only a list with top executives but also the personal address of CEO. This is why some members who use Ashley Madison have been subject to ransom demands. Ashley Madison site have been threatened with ransom and even a minor quantity of suicides were confirmed. This is perhaps the biggest and ever illustration of the fact that regardless of how vigilant you are of your online activity, you're online is something that you must always be aware of and be aware of the possibility that external influences can enter the picture and reveal personal information.

If you would like to look at some of the content found on the Dark Web then log onto TOR and type thehiddenwiki.org within the address field and then press enter. A huge list of websites that you can browse is displayed. I won't spoil the experience of you by divulging what's there , but I would advise you to be extremely cautious and use a high level of security measures against malware.

Based on what you plan to do and what you're feeling while visiting any of these websites it is suggested that you tape a masking tape on the webcam of your PC. If you do not the webcam could be watched by someone who is watching you and your house. This is certainly not in your best interests!

Chapter 6: How To Surf The Web Like A Hacker

Let me first explain the different ways hacking could take.

Hacking can be a grave crime and may be detrimental to people who aren't necessarily the intended victims in the incident. The results may manifest in a variety of ways , and can be devastating.

The most obvious way hackers could harm you is through identity theft. Identity theft is a major threat to those who suffer from it, and hackers are able to get hold of financial and identification information, and cause havoc to their victims' lives. In most cases hackers do this in order to get a motive. They might simply want to have fun with the information they've collected so they can buy items unauthorized. They can completely destroy your credit when they charge items to an existing credit card, and even apply for new credit cards, and then opening new accounts, and often seriously affecting their victims' financial situation. If the situation is ideal when the fraud is detected in the early stages, it could cause several months, or perhaps longer, of stress and effort to repair the circumstance. Sometimes, multiple fake IDs may have been created with the victim's personal information, and any activity that involves the IDs should be analyzed and followed.

The hacking of corporate and government websites can be catastrophic and sometimes result in a total shut down of a website until security issues or damage to the website has been identified and rectified. The harm caused by shutdowns could be lasting and cause massive financial losses. These shutdowns could also occur when the website is targeted by an attack known as a "denial of service" attack. Put in an easy way, the site has been targeted, and then bombarded by fake traffic, which results in the website not being able to process requests from legitimate traffic.

Hackers might also consider using malware and viruses to infect PCs. Most of the time, these programs are disguised by a beneficial program and that leads to the installing the harmful component at the same time. Certain software programs will proceed to make it appear as the appearance of a virus on computers to convince the user to buy fake antivirus protection or alternatively, the malware can be designed to track keystrokes to obtain passwords and other financial data. The malware can permit hackers to gain the control of a computer remotely and then execute nefarious tasks like a denial attack, and then make it appear as if it had been carried out by the affected computer.

An unnoticed information about hackers is the fact that none of them are malicious. Sometimes referred to as ethical hackers" or "white hackers" frequently assist governments as well as other potential target organizations to enhance their security. They also utilize every option that they can to stop massive security weaknesses that could be exploited. The "good" hackers are an important instrument in the battle against identity theft. They are working to limit the amount of cases.

If you search for "How to surf the internet like a hacker you'll find all kinds of details. Certain articles will provide suggestions for speeding up your internet browsing, while others will teach you how to protect yourself from hackers.

A hacker is typically an extremely clever individual who is trying to gain access to the computer networks and computers. There are many who would not like hackers to be able to do this, however in certain situations you can't stop hackers unless you have a highly effective security feature for your computer. Hackers want anonymity. They want the attention they get when they perform a task with success. They are generally considered to be unlawful, and should they be caught they could be liable to be punished with an enormous fine and a lengthy imprisonment. There are a few hackers that are not caught at all. They know the web as well as the back of their hand and are able to easily navigate through websites without being discovered.

Many hackers take what they do to challenge themselves. This is similar to the way people like solving Sudokus or crosswords. Similar to those who love these well-known games, more and more challenging hacks are needed by hackers who do this to have fun. They relish the thrill of being able to go into areas that normal people aren't able to access.

Some hackers may have criminal motives. You need to be able to defend yourself by installing malware or other protection on your PC. Whatever their motivation regardless of whether it's for intellectual curiosity or more sinister motives when a hacker gains access to your computer or network they could cause massive damage.

They could install programs known as Trojans or backdoors on the computers of their victims. Once they have done this, they will transmit information towards the attacker. They'll be able to access any details from your gadget in as little effort as a click. Hackers can work in pairs. They are compelled to work on their own because they only need to depend on them. At times, they may be part of collectives but you won't be able to find them frequently.

One of the most well-known hacker groups is known as Anonymous. Wikipedia has an interesting piece on Anonymous and I recommend that you go through it. The Scientology group suffered a lot from the actions of Anonymous and other organizations that have irritated Anonymous. In the aftermath of the Charlie Hebdo shootings in January 2015, Anonymous released a statement on Twitter condemning the shooting and declaring war against the terrorists behind the attack. He pledged to close any social media accounts associated with it. The report states that on January 15th they were able to shut down a site belonging to one of the groups believed to be to be responsible for the attack. however, critics have noted that by shutting the websites of extremists, you make it more difficult to trace the activities of extremists.

In this section, the need of hackers to be a secret was highlighted. Through the use in the TOR browser users can gain the anonymity hackers seek.

Chapter 7: Application

A growing number of people are trying to protect the inviolability of private life from specialized services that stick their fingers into other people's matters. More and more people want to be free of the "paternal concern" of government officials. state, and are seeking the right of the Constitution to decide on their own, independently, where to go and what to do what to do, where to look at and how to act.

The an anonymous network called Tor can be of assistance. Because it gives a human beings with a substantial diminution in the power of their persuasive attention, while eliminating all limitations when it comes to using the World Wide Web. Tor can hide your identity within the Network the entire thing you do on the Internet and on all websites you go to. It also allows you to navigate through your favorite websites with ease, which is carefully imposed to us by the governments we love that truly believe that they are more knowledgeable about us.

Additionally, the Tor network of Tor also has a minor practical advantage. It is often able to bypass these annoying issues, like a the ban of IP across different websites. These are nit-picky however they are very enjoyable.

People who are private utilize the browser of Tor particularly well-liked by those who seek to safeguard private information of their personal, as well as additionally, to safeguard access to blocked data. Because of the hidden services, users of Tor can build independent websites as well as other electronic resources as well as the exact location of servers are actually located and is concealed.

The web-browser that is part of Tor is often utilized by journalists for the purpose of interacting with informants in a safe manner. One of the most famous users to use this tool is Edward Snowden, transmitting with the aid of Tor various information to news media as well as Internet sources.

Non-governmental organisations' employees utilize the web browser of Tor to stay connected to special websites during their overseas business trips, without wishing to promote their work.

Tor is well-liked by civil rights activists from Fund of Electronic Borders, considering that this browser offers the ability to safeguard basic civil rights and freedoms within the world's network. Many corporations use Tor to secure analysis of the work of their rivals in the marketplace. The web-browser from Tor is utilized by various special services to ensure secrecy in the performing specific jobs.

The structure and the principles of work.

Connections to outgoing connections anonymous

What exactly is this an anonymous network like Tor? Tor can be described as an abbreviation for "The Onion Router". If you are fascinated by boring technical details go to the page of Tor on Wikipedia and read about it. If you'd like to make it simpler, look up the exact identical page on Lurk further. However, I attempt to make it clearer.

While this network works as a basis for the normal Internet however, information does not transfer directly from your computer to the server, and then back like a "big" network. Instead, the information is pushed through a long line of servers that are special and ciphered multiple times throughout each step. In the end, the final user, which is you, is completely anonymous to the websites and your actual address, it is displayed as completely incorrectly, without having any relationship to you. Your entire movement cannot be tracked, and neither can the actions you took. Also, intercepting your activities is unusable too.

It's the theory. In reality, everything is not as optimistic. We'll talk about the various possibilities in the near future. It's been a lengthy and boring introduction isn't it? Are you feeling a little agitated to setup and try to utilize this amazing technology? So, let's start!

The system that is used by Browser Tor lets its users to connect to their own computers, which are known as "Onion" Proxy-servers, which then later connect directly to main

Tor servers, which are responsible for organizing Tor web-chains (they utilize multilevel programming). All data transmissions that go through the system are routed through three separate proxy servers, and their choice results in accidental.

Before sending out a packet it is encoded using three keys. The first network pack receives the data packet, then encodes its "top" levels of the codes (similar as peeling an onion) and determines which location to forward the data packet. The other two networks do similar things.

In the inner Tor networks, traffic is directed between routers, before finally reaching the point of output, which is where the encoded data has already reached home server. Then, the traffic from the recipient returns to the original Tor point in the network.

Anonymous hidden services

In 2004, Tor began to make servers invisible, hiding their location within the World Wide Web using special methods for anonymous networks. You can access certain services that are hidden only by using Tor client.

Access to hidden services may be gained by using specific pseudo-domains that are part of the top-level ".onion". Tor networks can identify them as anonymous and forward the data to hidden services that are specially designed for this purpose. They process data with normal software that is tuned for the proper listening to closed interfaces. The"domain ".onion address" are generated by the server key that is opened and consist of 16 numbers as well as Latin letters.

Restrictions

Tor is designed to conceal clients' connection to servers. However, complete confidentiality cannot be achieved since coding here is only a means to achieve anonymity on the Internet. To achieve a greater level of privacy, it's essential to have additional hardware security. It is also recommended to utilize stenography when recording data.

The basic advantages of Tor Browser

Tor browser offers these advantages:

Access to ANY website from ANY region of the Earth regardless of the service provider you choose;

Tor browser alters the IP address of the client to ensure complete privacy. 100% guaranteed.

The browser is easy to install, and its use is completely free.

Repeater networks can be used in conjunction with them;

Security from web tailing threats privacy of data;

Security-related threats are automatically blocked.

The protection packet is not installed. It's launched from all devices, even portable.

The fundamental negatives of Tor

The Tor Browser comes with a few downsides:

Too slow loading speed

The videos may not all be played.

Very low security.

The setup of Tor

Tor for Windows. Installation from the Tor Browser Bundle.

Open any browser (Mozilla Firefox, Internet Explorer or other) and enter in an address line: https://www.torproject.org/projects/torbrowser.html.en. If you locate the Tor Browser Bundle using the aid of the search engine be sure to verify that the address is correct.

Press the big purple button "DOWNLOAD" to create the file that will be installed by the program. Tor Browser Bundle.

A web page will establish the operating system of your computer automatically. loading of the required file will start. If, for any reason, you need to download the file from installation to another operating system, then you can select the necessary version from a selection.

Most browsers will require confirmation that you want to download an image. Internet Explorer 11 displays the field with an orange frame on the lower left-hand the window in the browser.

It is suggested to save the file on an external disk, independent of your web browser. Press the button "Save". This is the display of what the application is called Tor Browser Bundle version 5.0.4 that was in use at the time of the writing of this article. The latest version of the software is available.

Configuration of the Tor Browser Bundle

After the loading is complete, you might be prompted to open a folder in which the files were stored. It is by default an unnamed folder called "Downloads". Start the file torbrowser-install-3.6.2 en-US.exe by a double click.

If you double click the installation file, the program will open with a warning about the source and the source. Always consider these warnings seriously. It is crucial to be sure you be confident in the software you have purchased and that you have the genuine copy from a legitimate website that is a secure communication channel. It is clear in this instance that you will need to know where to download the software. Downloading was made available on the secure HTTPS website of the Tor. Tor. Click "Run".

The window that you choose for the language used by the Tor Browser Bundle will open. Select a language from many variants and then click "OK".

In the next window, it is recommended to select the folder to set up Tor Browser Bundle. The default setting is a desktop shown. You can change the setting location however we can keep an address unchanged.

There will be an update window indicating the success of the setting. Click "Finish". Tor Browser will begin immediately. Although it is not a perfect mark to the direction at "Run The Tor browser bundle". We'll return to using Tor Browser Bundle following some time. If you did not clean an area and the program Tor Browser was started, just close the window.

The Tor Browser Bundle is not set within the system, like other programs are, and it will not appear in the list "Starting" in your PC.

Use of the Tor Browser Bundle

First time to start Tor Browser

After setting up, the settings, we chose not to launch Tor Browser, therefore now you'll be able to begin the program for the very first time. If you have followed the instructions when setting up, you'll see in your computer a file titled "Tor Browser".

Navigate to the directory "Tor Browser" Double-clicking will open"Start Tor Browser" "Start Tor Browser".

When you first launch Tor Browser the user will be able to see an interface that will inform you to choose to alter the tuning. You might need to revisit them later. During the meantime, trying connecting to the network of Tor using"Connect" "Connect".

Then a new window will be opened with a green field. At the beginning of Tor will remain open slightly longer.

In the initial launch of Tor Browser it may need just a little longer than usual, however, be patient. Within a few minutes Tor Browser will tune connection. A web-browser will be displayed that will be happy with a an excellent start.

Tor for Ubuntu

To install Tor Browser in Ubuntu is to download it from the official website. It's the only correct and logical method.

The best and most reliable method to set to set up Tor Browser is to download it from the official site

https://www.torproject.org/download/download-easy.html.en

Select of the software that conforms to the architectural system, select English, and download:

Unpack downloaded achievements in the home catalog, and then transfer it to it: you will find the following executable file

We reveal the authorization for execution within the properties

This is it; simply double-click this file will start Tor Browser

If, after double-clicking Tor Browser does not open but instead, a text editor, it opens you must allow the scripts to be executed in the settings of the file manager. Nautilus:

If you'd like to put an appendix with a label within Dash's main Dash menu,

You can read it on it. Internet it.

Start Tor Browser we move to the website for checking IP.

If everything is right, we'll see something similar to this

Where do I come from? Naturally, I am from Germany And, more importantly the operating platform is "Windows"

Installing of Tor Browser in Ubuntu 14.04-12.04 from the repository

Install option from a repository that is not the latest version: In order to install the Tor Browser Bundle on Ubuntu start the terminal and perform the following steps based on your system

For Ubuntu 32-bit:sudo add-apt-repository ppa:upubuntu-com/tor

sudo apt-get update

sudo apt-get install tor browser

sudo chown $USER -Rv /usr/bin/tor-browser/

For Ubuntu 64-bit:sudo add-apt-repository ppa:upubuntu-com/tor64

sudo apt-get update

sudo apt-get install tor browser

sudo chown $USER -Rv /usr/bin/tor-browser/

It's that easy, the software is now installed. You will be able to locate it with the aid of the menu Dash

Other languages available in the Tor Browser Bundle, when installation comes from repository

Tor Browser is Firefox of stable version. We will modify different languages.

Take out the Help-About section of the insert. Tor Browser

Take a look at the browser version and proceed to the site using other languages.

Mozilla-Firefox

Choose the browser that is compatible with your version. download the package in other languages and install it.

Enter the address in the line of address

about:config

We agree that we'll be cautious "I'll be vigilant I swear!"

Enter the search phrase into the search box

general.useragent.locale

Change the meaning of this parameter to the en-Us to de (fr)

All that's left is you can start Tor Browser.

Now , it's possible to be anonymous "to roam around the networks"

Switch to Flash Plugin and JavaScript in Tor Browser

If you'd like to view flash-based movies using this browser, it's simple to enable it. Additionally, it is possible to permit the execution of scripts. In this case, however, security is reduced!

I would not recommend it if you are a confirmed paranoiac.

Also, if you're looking for flash to begin working

Change to "Tools" Move in "Tools "Additions"

The insert "Plug-ins" toggle to Shockwave flash.

Now, click to your preferred website and enjoy online videos, for example, on YouTube.

In the area "Expansions" there is a way to disable the expansion that stops scripts from being executed on websites:

This is all there is to it, it is possible to view flash videos and scripts that will be implemented.

This all breaks security, to ensure that we have installed with this program.

Tor for Mac

Acceptance for Tor Browser Bundle

Open any browser (Mozilla Firefox, Safari or other) and enter in an address line: https://www.torproject.org/projects/torbrowser.html.en. If you discover the Tor Browser Bundle using the aid of the search system then you can be sure that the address is correct. found address.

Press the big purple button "DOWNLOAD" to start the installation file of the program Tor Browser Bundle.

The website will automatically define your operating system, and the loading of the needed file will start. If you need to install the installation file on a different operating system, you are able to select the appropriate version from the options.

If you're using Safari then the download from Tor Browser Bundle will begin. If you are using Firefox you'll be asked to save or open the file. It is always recommended to save your file, this is why you should click "Save". In this case Tor Browser Bundle Version 4.0.8 is available, and is current at the time of the publication of this guideline. As of the time of reading, perhaps, a an updated version of the program is expected to be available.

Configuration of the Tor Browser Bundle

When you have finished downloading, perhaps you'll be prompted to open the folder that the your file was saved in. In default, it's the folder called "Downloads". Launch the program Tor browser 4.0.8 4.0.8 - - osx32_en US.dmg US.dmg with a double click.

A pop-up window will show advising to install Tor the Browser Bundle by drag the program to the application folder. Do it.

The application Tor Browser is set in the application folder.

Utilization of the Tor Browser Bundle

To launch Tor Browser in the first time, search for the application on Finder or (in other update versions of OS X) in Launchpad.

When you click at the symbol in Tor Browser a window will show warning about the source and the source. It is important to be aware of such warnings. It is crucial to make sure that you believe the program, obtained a genuine copy from a legitimate website that provides an encrypted communication channel. In this instance, you know the requirements and how to obtain the software. Downloading was performed through the secure HTTPS website of the Tor. Tor. Click "Open".

On the initial launch of Tor Browser the user will be able to see an interface that will inform users to alter the settings. You might need to revisit it later, but at the same time, you're trying to connect to the Tor network by Tor by pressing"Connect "Connect".

A new window will be opened with a green background. The window will be opened at the beginning of Tor will remain open for slightly longer.

In the beginning of Tor Browser it could take longer than usual however, be patient. Within a couple of seconds Tor Browser will connect. A web-browser will be displayed that will be happy with your an excellent start.

You can check, whether you are connected to the network of Tor, visiting check.torproject.org. If you're connected, the website will say: "Congratulations. This browser is set to utilize Tor".

The web surfing experience through the network of Tor differs from regular work on the Internet. We advise you to adhere to these guidelines for safe surfing the internet via Tor and to ensure your privacy.

Now , you're ready to start enjoy your anonymous surfing on the Tor network. Tor.

Configuring and running bridge mode

Installation of Tor with bridge/relay in the bridge/relay configuration.

Installation is very easy - you just need to install the software, and then run the installation.

There are two forms of distribution including The Tor Browser Bundle and Vidalia Bridge Bundle. Tor Browser Bundle is aimed specifically at safe browsing on the Web. Vidalia Bridge Bundle lets you users to safely browse to the Web but also expands Tor network with your personal computer.

1. Warning of inability to run Tor Bridge service

[Warning] Unable to connect to 0.0.0.0:443 address that is already used [WSAEADDRINUSEWSAEADDRINUSE.

Is Tor already in operation?

The reason for this is that the identical port on the same computer was utilized by Skype. This issue can be fixed by following the steps: Vidalia Control Panel -Settings> Sharing Basic Settingsthe Relay Port section: Here you need to change 443 to another number, like 4444 (This port wasn't utilized in any other software)

2. Warning of files from GEOIP' inadequacy:

[Warning] Failed to open GEOIP file C:\Documents and Settings\User\Application

Data\tor\geoip. ...

[Warning] Failed to open GEOIP file C:\Documents and Settings\User\Application

Data\tor\geoip6. ...

The problem is that geoip6 and geoip files suddenly appeared in another directories, such as C:Documents and SettingsUserL ApplicationsSettingsTor. The issue can be fixed by simply copying the files to the right directory.

3. Warning of inability to connect to the bridge server via external sources:

[WarningWarning! Your server (aa.bb.cc.dd:4444) hasn't been able to verify that its ORPort is accessible. Check your firewalls port address, /etc/hosts and so on.

The reason for this is because D-Link router is able to provide Internet connection through NAT. To make port 4444 accessible to the outside world via global IP aa.bb.cc.dd and aa.bb.cc.dd, you need to set port forwarding on the LAN out.

Tor-D-Link-port-forwarding.

4. Note that your contact information isn't set.

Note: The ContactInfo configuration option isn't configured. It is recommended to set it to allow us to contact you in the event that your server is not configured correctly or something else is wrong.

It is not necessary to fill in your contact details, however you are able to do it. This can be done through the Vidalia Control Panel Settings> Sharing Basic SettingsHere you can fill in your Nickname as well as Contact Information (your email address).

5. Beware of making timers that are set to the "wrong" date:

[Warning] Received directory with skewed time (server '82.94.251.203:443'):

It is possible the clock in our house is ahead of 56 mins, 7 seconds or that their clock is behind. Tor requires an accurate timer for operation: make sure you check your time, timezone, and

and time settings for date and time.

The time is oddly different during an hour (my clock is pushed ahead for 56 mins) It's as if the issue occurs in winter/summer time. The cause is a glitch in the Tor server. What can be done to fix it?

1. Run the Tor system and then wait for it to load completely (the moment that Tor establishes the connection and the onion icon on the tray changes to green)

2. Set the date and time settings as well as set it one hour later or earlier. The connection will eventually be lost, however it will be restored in a certain time.

3. It should take approximately 15 minutes, then bring the time back. The connection may be lost and again, but Tor will be restored to regular mode.

The proxy mode can be tuned by tuning it.

How do you configure proxy settings in Internet Explorer.

For OS that runs Windows 7 it is necessary to open Control Panel, and then go to Properties of the browser. You can then go to the Connecting inset, which is in the lower right corner, click on Tuning of the network. It is necessary to label the area "Proxy-server" after which you can open "In addition" in order to enter an inset you will see digital values displayed on a photograph.

Explorer operates through Tor.

How do you configure proxy settings in Google Chrome.

It is first necessary that it first be firstly the "default default browser" on your PC. Further:

Click on "Change the configuration of proxy-server". Insert to tune Internet Explorer should be removed (see the image).

How do I configure proxies within Opera

It is essential to enter "settings" and then propose these in line with the version of the browser you are using. We will mark the area of Socks in addition, it is required to enter subsequent numerical data: 127.0.0.1: 9050

How to set up proxies within Mozilla Firefox.

The user needs this plugin (https://addons.mozilla.org/en/firefox/addo n/foxyproxy-basic/?src=search). It is to be configured within "Expansions".

After you have set up the plugin, you must select "Tor proxy across all addresses".

In addition, you must suggest the setting.

There should not be "forbidden websites" today.

If you want to know IP-address, appeal to http://www.checkip.com (ip-check.info)

If a user is working via Tor it is possible to have an address different from the one used by the service provider.

How do I "TORify" How to "TORify" ICQ Skype, uTorrent.

The procedure is similar to ICQ or Skype: "Tools -- Settings -- Also Connections" Find the entry SOCKs5 then write the following numbers: 127.0.0.1:9050

For uTorrent you should move into "Settings-Settings of the program-Connections". You should then select the settings that are shown in the image below.

Relay mode

The safety and effectiveness of the Tor network are dependent upon the amount of Nodes which are reliable in sending traffic. These are referred to as relay nodes. The EFF even organized Tor Challenge in order to encourage as many users as possible to create and configure these nodes. In actual this article is dedicated to this easy task. To work in relay mode, you'll require a server which is where Tor Relay will work. You can connect to your home computer or could reconfigure a router that is smart. There is another option - to make use of VPS (Virtual Private Server). Tor software is fairly simple and is able to work with VPS with only a few configurations. Memory of 256MB or even 128 MB is sufficient. Disk requirements are not too high either that is, less than 1GB. The cost of this monthly server is comparable to the cost of a cup of coffee.

Therefore, we can create a VPS. It must be verified as an outside IP. For me I am happy with your server, however there are many VPS's with Linux or *BSD installed on them. In general, when purchasing , you receive an already installed Linux distribution. Pick whichever you prefer. I will demonstrate using Debian to show an instance.

In the beginning you must start by installing Tor to your server.

aptitude Install tor

In default Tor will operate in mode of a web client. You are able to use it online, however for all other users it's not useful. The traffic of someone else's won't flow through it. You have to enable Tor Relay mode.

You must also switch on Directory Service &mdsah catalogue service which is reliable in spreading info about different Tor servers. It is possible to use unspecified ports to send and catalogue. The default configuration file will use port 9001 to retransmit packets and broadcasting, while port 9030 is for catalog service. However, we will allow our server to be accessible for port 443 as well as 80. These ports are typically used to handle web traffic.

Write in the following lines:

Nickname MyCoolNick

Contact Information Person

ORPort 443 NoListen

ORPort 9001 NoAdvertise

DirPort 80 NoListen

DirPort 9030 NoAdvertise

ExitPolicy reject *:* # no exits allowed

ExitPolicy reject6 *:* # no exits allowed

Under'Nickname', type your server's name. Then you'll use it to control of server activities using special services available on TorProject.

In the Contact Info line you can add your contact details (in the event that someone should need to get in touch to you). If you don't remove it, but the server won't be able to let anyone know who the owner is.

The two lines in the last paragraph prohibit the use of our server as an the Exit Nod to traffic. If not, Tor will try to utilize our server for transfer of outgoing traffic from the networks that use external servers. However, not all users use Tor in good faith and if traffic is diverted from Tor via your server it may affect you.

Furthermore, the configuration makes a server inform other users of the network that it is available on ports 443 to send packages, and ports 80 for the announcement of information regarding other servers on the network. In reality, a server must be waiting for reports on ports 9001 or 9030. In Debian

Tor by default does not work in a tunnel, and this configuration can prevent problems when connections to ports.

Through iptables, we will be able to influence the necessary connection between ports right now.

If there is a particular tools to tune the Iptables' network screen in the distribution you have chosen It is possible to make use of it. It's easier and clearer to make everything you need yourself.

We create the file of /etc/iptables.save.rules of such content:

Generated by iptables save v1.4.14 on Sat 5 Jul 14:15:04 in 2014

*filter

INPUT ACCEPT [0:0]

FORWARD ACCEPT [0,0]

Output ACCEPT [22:1968]

A INPUT -m state --state RELATED

-A INPUT -i lo -j ACCEPT

A INPUT -D 127.0.0.0/8 ! -i lo -j REJECT --reject-with icmp-port-unreachable

-A INPUT -p tcp -m tcp --dport 22 -j ACCEPT

-A INPUT -p tcp -m tcp --dport 80 -j ACCEPT

-A INPUT -p tcp -m tcp --dport 443 -j ACCEPT

-A INPUT -p tcp -m tcp --dport 9001 -j ACCEPT

-A INPUT -p tcp -m tcp --dport 9030 -j ACCEPT

-A INPUT -j REJECT --reject-with icmp-port-unreachable

Commit

Finalized on Sat July 5 at 14:15, 2014

Generated by iptables save v1.4.14 on Sat July 5 at 14:15, 2014

*nat

PREROUTING ACCEPT [0:0]

INPUT ACCEPT [0:0]

Output Acceptance [1:104]

POSTROUTING ACCEPT [1:104]

A PREROUTING -p tcp TCP --dport 443 Redirect to ports to ports

A PREROUTING -p tcp TCP --dports 80 -j REDIRECT to-ports 9030

Commit

Finalized on Sat Jul 5 14:15:04 , 2014

This allows us to improve the efficiency of our tor server as well as access to ssh to allow remote administration.

It is up to the administrator to determine the how to load these rules. Usually I prescribe the start of iptables - restore in /etc/network/interfaces:

auto lo

iface lo inet loopback

pre-up   /sbin/iptables-restore
/etc/iptables.save.rules

On Your server the file of /etc/network/interfaces is being rewritten each time at re-starts, it is therefore possible to do hardly differently.

For example, you can put the iptables loading rules into /etc/rc.local. In EOF before exit 0 , we add the line.

/sbin/iptables-restore
/etc/iptables.save.rules

In the end, we reboot tor server:

Service tor restart

We verify that we have done everything correctly. After a short time, following the restart of the file /var/log/tor/log lines should be visible:

Self-testing confirms that your ORPort is accessible via the outside. Excellent. Publishing server descriptor.

Tor has been able to open the circuit. It appears like the client's function is functioning.

Self-testing confirms that your DirPort is accessible via the outside. Excellent.

Conducting a the bandwidth self-test...done.

In hour or two, when information will revive in a database, it is possible to call on globe.torproject.org/ and, writing nickname of the server in the line of search, to make sure that the network of Tor was filled up by another point of redistribution of data.

The first step is to change the server so that the traffic won't go. The course of life that is Tor Relay is a theme of a separate article.

UPD: As with distributives, the most recent version of Tor isn't always the truth it makes sense to link specific repositories.

Also, for Debian as well as Ubuntu it is possible to connect to the official repository of torproject.org. For this purpose in /etc/apt/sources.list.d/ we create the file of torproject.list of next contain:

deb http://deb.torproject.org/torproject.org DISTRIBUTION main

In place of DISTRIBUTION we create the version of your distribution (for instance jessie or saucy) Do it

gpg --keyserver keys.gnupg.net --recv 886DDD89

gpg --export A3C4F0F979CAA22CDBA8F512EE8CBC9E88 6DDD89 | apt-key add -

Update to apt-get

apt-get install tor

* tor

* ,vps

*,tor relay

Chapter 8: Adjustment, And Work Using Vidalia Polipo Shell Vidalia Polipo Shell

There are Internet providers that ban users from using Tor. Repeaters must assist locked users using Tor gain access. Since bridges aren't listed in the directories of public domain as normal repeaters, then the they are not able to block access to bridges in all locations. Open addresses of bridges can be found here https://bridges.torproject.org. Or one can write a letter to bridges@torproject.org. Indicate subject "get bridges". Requests should only be made via an account Gmail.

It is important to realize that the reality of Tor installation is that it does not encrypt the network connection of your computer. Other software components and adjustments are needed. Software program Tor solely controls cyphering, and determines how software suit's movement over the repeater networks.

1. In the beginning, we require a proxy servers installed on the computer of the user. Sometimes, it's referred to as "filtering

proxy". It is a proxy between applications used by users within Tor network and Internet as well as the Tor network.

There are two fundamental models of proxy server filtering two basic versions - Privoxy as well as Polipo.

Some years ago, developers of the Tor system advised the use of Privoxy. In every assembly only Polipo made available on torproject.org. (?)

It's difficult to evaluate the two species based on their features. Polipo is considered small - smaller than 200K. The entire adjustments it makes are stored inside the file polipo.conf. I haven't been able to find comprehensive information on the configurations. Maybe it's not necessary.

For use using the Tor system, one must use a the polipo proxy version that is not less than 1.0.4 as earlier versions don't allow the use of SOCKS protocol which is why they are not compatible with the Tor system.

Privoxy is a no-cost web-proxy offering improved capabilities for filtering Internet content to ensure the security of Internet users and privacy protection. The most recent version was 3.0.17. (2011). But Privoxy is frequently used to act as an intermediary between applications and software programs Tor. It is worth noting that Privoxy can be an independent program, which is able to protect the user interests on the protocol HTTP.

The proxy that you will use on your computer is an individual decision. It is highly unadvisable to combine them because both proxy servers utilize the port 8118. Moreover, in the event of a combined operation, issues can occur.

The most straightforward advice is For those who don't wish to deal with the hassle, it's recommended to utilize Polipo which is one of the final assemblies on the site torproject.org. Anyone who wants additional features to make adjustments, download and install Privoxy before the installation process, exclude Polipo out of installation.

2. To manage Tor system load and work management, the program Vidalia is

employed. It is commonly referred to as a graphic shell to describe Tor.

The settings for Vidalia there is the option to start Tor as well as filter proxy at the launch of Vidalia as well as stop and start Tor while in operation, browse through Tor network map , and many more. The work with Vidalia will be covered in detail. Tor parameters settings can be changed using the Vidalia shell.

When you launch the software Vidalia is a symbol Vidalia appears as an onion. When operating systems are installed, Windows it is located in the tray of the system (near it is the wristwatch, check the image). For the operating system Ubuntu it is displayed in the taskbar. You can open the video window Vidalia by pressing a left mouse button on its icon.

1 2

On the first photo Tor is off, while the second picture, it is switched on.

If you've installed Tor the proxy filtering server as well as Vidalia, you can modify applications to use using Tor or, as they say "to charge applications".

The installation of Tor onto Windows Operating System -- Vidalia Bundle pack

In contrast to Tor Browser all the other assemblies (packs) execute installing Tor and other components.

Components function in the exact similar to the Tor Browser, but there are some finer aspects. If, for instance, you are using the Firefox browser, if Mozilla Firefox isn't configured, the TorButton won't be configured either. That's why it's recommended to setup Firefox prior to the installation of Vidalia Bundle. Vidalia Bundle.

The following images show the Vidalia Bundle installation procedure for Windows 7 >:

Select the load file, and secure it

Start the setup file

The necessary options are marked by"by default" ticks "by default"

If a user wants to switch to a different configuration, like to use a filtering proxy server Privoxy or any other browser to perform anonymity, ticks need been removed from unneeded components. Also, the browser and Privoxy should be setup prior to that.

In earlier versions there is a second option:

Assembling The Vidalia Bundle to Windows includes Tor, Vidalia, Polipo and in earlier versions , Torbutton. Torbutton (the variety of different versions is shown on the photos).

If Firefox isn't installed on a PC, the software installation program alerts about it, and advises you to install it again and then install.

All of the standard configuration elements are default to be adjusted to allow joint operation.

Then, select the load directory, or you can leave the suggestions:

View of windows for setup

The program Tor is configured as a client-side application by default. It is built-in with a configuration file which is used by most users don't need to alter the settings.

Tor parameter settings can be changed with the aid of the Vidalia shell

The program software Vidalia functions as a graphic Shell for the Tor system. It is compatible with all platforms , including Windows, Mac OS, Linux and many other Unix systems.

In the event that it is the Tor Browser assembly is used then Vidalia is launched using the file Start to Browser.exe from the catalog

If the Vidalia Bundle pack is selected, you can launch this file vidalia.exe from the catalog:

In the event of launching a sign Vidalia will appear as an onion. When using Ubuntu, under the Ubuntu operating system, it is displayed in the taskbar. On Microsoft's Windows operating system, it is in the tray for the system (near that watch).

To launch "Vidalia Control Panel" you need to click a left mouse button located on its icon.

Vidalia settings are easy to understand and understandable. Although we only briefly mention the following:

-- Tor Start/Stop (Start/Stop Tor)

Settings for servers (Sharing) create the operating modes (client bridge, server or client)

Overview of Networks (Network Map)

Displaying Tor the network maps:

In Tor regular operation the circuits that are in use must be displayed on the bottom of the window. Also, in the adjacent window to the right, the servers of the circuit as well as their specifics must be displayed. In the upper window , their physical location is shown.

With the aid of the network map , you can select servers based on their location or speed.

Modify the Identity (New Identification). It alters Tor circuit, and the result is that it outputs IP-address.

After the successful transfer of tray, a message will be displayed.

- Traffic schedule

It displays output and input traffic as well as Tor rates:

the Message Log. It lets you see Tor operations logs

- Settings. The window opens "Settings":

- - Flap "General" allows the setting up of Tor components' launch procedures

- - Flap "Network" allows the writing out of proxy servers ("I make use of proxy for connection to Internet") or bridge ("My provider blocks an internet connection through the Tor network") (read in the Internet and in the Internet). Tor blocking section and how to handle it.

"Advanced" – the flap "Advanced" allows the setting of (checking) the parameters for TCP connections (127.0.0.1 port 9051) and also determining (controlling) the location of torrc settings files and data catalog. Additionally, you can modify the configuration file.

-Flap "Appearance" allows you to change views based on your system

- - Flap "Services" allows the addition of addresses and ports for computers on the Vidalia network.

- The Flap "Help" is a call to"Help", which calls the Vidalia Help Desk.

As you will see from the previous paragraphs using the Vidalia shell, you can configure and manage quite a few Tor the system's parameters.

Tor delicate adjustment

In general, the default settings, as implemented by the Vidalia shell, are enough to allow full value anonymous use on the Internet. In some instances, however, you might require further modifications to Tor parameters.

The changes are implemented through editing the Tor configuration files. They are referred to as delicate adjustment.

Filtering proxy Polipo configuration file

Here is the most basic version that is a simpler version of polipo.conf structure file can be shown (only not the commented-out instructions).

```
# Basic configuration

proxyaddress is "127.0.0.1"

proxyport = 8118

allowedclients equals 127.0.0.1

permittedports = permittedports =

proxyName = "localhost"

cacheIsShared = false

socksParentProxy = "localhost:9050"

socksProxyType = socks5

chunkHighMark = 33554432

"" diskCacheRoot= ""

disableLocalInterface = true

disableConfiguration = true

DNSUseGethostbyname = Yes

disableVia = true
```

censoredHeaders = from,accept-language,x-pad,link

censorReferer = maybe

maxConnectionAge = 5m

maxConnectionRequests = 120

ServerMaxSlots = 8

ServerSlots = 2

tunnelAllowedPorts = 1 - 1-65535

Configuration file is a standard text file. It's name is the torrc (with the extension '.') it is found in

If you are making use of Tor Browser assembly - in catalog..It is possible to use the catalog. \Data\Tor

- in installation packs - \Application Data\Vidalia

- within the Ubuntu Linux operating system - in the catalog the directory /etc/tor

Software program Tor when load (reloading) first takes a look at the configuration files and then sets up operating characteristics in line to the values in the torrc file.

Torrc editing of files can be done using an elementary text editor like Notepad, AkePad etc. It is suggested that prior to editing, you backup the your original torrc file within this same location. For instance, you can add to an existing name the extension *.bak, *.001 etc.

To take effect the changes, you must reload the entire Tor System software!

1. Correction of input or output Node in the Tor network

Interaction with Tor users reveals such a subtlety - not everyone is happy with the constant changing the IP address of their computer.

It is important to remember that the output servers of Tor always change in a random fashion. A user's IP is a sign that their IP is in unstable. Regarding the resource that is attended, the user could change from being an American or an American or Frenchman into , call it Japanese, Hindu or any other type of person.

This approach basically improves the privacy but it can be not acceptable (for instance , when working with websites that can fix the session of users).

In Tor there is the possibility to specify directly which server is to be used for output. In this scenario, the IP will remain constant. Tor creators don't recommend doing this as it can compromise anonymity. The user is required to choose on their own what is most important, but I'll tell you ways to stop an ongoing change in IP.

You will have to edit Tor configuration file, it is called "torrc" and you can get to it either through "Start" -> "Programs" -> "Vidalia Bundle" -> "Tor", -> "torrc", or find in the folder \Documents and Settings\user\Application Data\Vidalia orrc. Torrc A text file commonly used It is opened using the notepad.

In the torrc, write two lines:

ExitNodes

StrictExitNodes 1

Where:

Variable ExitNodes: indicates that you want that you will use a specific server as the output node

StrictExitNodes 1is a sign that in the event of an outage of the server you choose, you don't attempt to connect to another server, but instead make the error.

It is possible to write multiple nodes separated with commas or in the case of using ExitNodes de, that way, we'll get only German servers for outgoing servers ("turn into" the status of a German!).

You can find necessary server at: http://torstatus.kgprog.com/ or https://torstat.xenobite.eu/

These are the listings of Tor server networks, and you can choose the one that is needed in accordance with the speed, country, and traffic. Attention should be paid to the ability of a server as output.

It is evident that servers that aren't output servers, can't perform this function.

Choose a server , and then write the name (Router Name or Nickname) such as:

ExitNodes 1000rpmLinux

StrictExitNodes 1

Changes to the configuration file are safe and that's it, the configuration file IP is fixed. It is also permitted to create multiple nicknames separated by the commas (nickname1 nickname1, nickname2 Nickname3, nickname3) In this scenario, output servers will change in real time but they can be picked from the allowed.

It is evident that the efficiency of the network in this scenario is contingent on access to an output server and in the event Tor ceases to connect to websites at first, you must determine if the output server has gone out of whack.

The input node too is fixed:

EntryNodes

StrictEntryNodes 1

Another setting that changes the the output of nodes (host) for specific domains that allow you to protect sessions for servers that monitor IP clients. The syntax of record is as follows: following:

TrackHostExits host,.domain ,...

2. Exclusion of nodes that are suspicious

To exclude nodes that are not trusted (for instance - Russian, Ukrainian, Turkish) you must include in torrc ExcludeNodes: ru, ua, tr

You could also provide the names of a specific list.

If curious people with gray eyes in these countries have the idea of creating an untrue Tor-server and try to access output data, then we will not be able to connect to the service in any form.

There is a great feature of the torrc file. It is a commentary. Tor will not execute an entry in torrc in the event that a line begins with a sign "#". Thanks to commentaries, you are able to store your data in a safe torrc format and, if required, switch them off by removing "#".

3. Writing a proxy-server for Tor

Include the following lines at the close of the Tor configuration file, changing and (as well as and If they're) in particular values of connected http or https proxy-server.

Require Tor to handle the majority of HTTP directory requests via this host:port (or

host:80 (if port is unset).

HttpProxy :

A username-password pair that can be used in conjunction with HTTPProxy.

HttpProxyAuthenticator :

Forcing Tor to establish the entire TLS (SSL) connection via this host:port (or

host:80 (if port is unset).

HttpsProxy :

A username/password combination that can be used in conjunction with HTTPSProxy.

HttpsProxyAuthenticator   <   login   >:< password >

After you have corrected and saved the torrc files, you must restart Tor.

To check settings you can use Vidalia graphical shell or Tor-analyzer (go to http://check.torproject.org).

The list of different Tor directions (settings)

EntryNodes nickname,nickname,...

This is an inventory of servers that are suitable to use as an "input" to establish TCP/IP-connections using Tor routers' nodal circuit in the event that it is feasible.

ExitNodes nickname,nickname,...

It's a list servers, that will ideally play the function of closing links in Tor routers' nodal circuit in the event that it is feasible.

ExcludeNodes nickname,nickname,...

It's a list that lists nodes that should not be used for creating a nodal circuits at all.

StrictExitNodes 0 If they are set up in the 1st position, Tor will not utilize any nodes other than those that are listed included in the listing of nodes that output to act as mediators establishing connection with the the host of the target and are unique closing links to the circuit of nodes.

StrictEntryNodes 0

When the parameter value is given to this parameter, then Tor does not make use of any type of nodes, except the ones that are included on the input list used for Tor network connections.

FascistFirewall 1 - 0

If 1 value has been associated with this parameter, then Tor when creating connections will be referring exclusively to Onion Routers which have strictly precise port numbers (with which your firewall allows the establishment of connection) available to establish connection (by default the port number is 80-th (http) and 443-rd (https) See FirewallPorts). This will permit Tor that is running on your system to function as a client for a firewalls that have strict policy for limiting access. The opposite is not true because in this scenario Tor can't fulfill the role of a server shut down by a firewall.

FirewallPorts Ports

A list of the ports that your firewall will allow connections. This is only used in conjunction with the an adjusted parameters of FascistFirewall. (by default: 80, 443) (Default: 80, 443)

Ports with a Long-Lived Term

A list of port numbers available for services that are prone to establish extremely long connections (among them are chats and interactive shells). Nodal circuits that originate from Tor routers, that use the ports listed above, have only the nodes that have the highest uptime (typical time of being present in the network) in the hope of making it less likely for disconnected nodal servers from Tor network prior to the end of the flow (by default 21-22, 706 1863, 5050, 5190, 5222 and 5223, 8300, 6667).

MapAddress address: new_address

If a request for the specified address is sent to Tor the onion router, it changes address before taking on the processing of the request. For example, if you want Tor nodes circuit to be used during connection to www.indymedia.org with output through torserver (where torserver - is a pseudonym of server), use "MapAddress

www.indymedia.org
www.indymedia.org.torserver.exit".

NewCircuitPeriod NUMBER

Every number of seconds to evaluate the status of the connection and make the decision on whether the formation of a new nodal circuit is required to be started (by default 30-seconds).

MaxCircuitDirtiness NUMBER

to allow a repeatable use of circuits that was to allow the very first time that it in a specific combination of its links, the largest number of seconds ago, however never join a flow of a circuit that has been in use for quite an extended period of period of time (by default 10-minutes).

NodeFamily pseudonym,pseudonym,...

The denominated Tor servers (in the same way in order to improve the degree of transparency Tor network structure) form an "family" in the context of general or joint management and therefore you should be cautious about using more than two of these nodes "related to family bonds" in one chain anonyme Tor routers. A special function of the option NodeFamily may be required only when the server that is referred to as NodeFamily is not able to identify the "family" it believes it belongs to it should be announced by indicating that parameter MyFamily in the torrc file located on the server's side. OR server. Multiple options for this option are possible.

RendNodes pseudonym,pseudonym,...

This list contains nodes to serve as points of rendezvous (meeting) in the extent feasible.

RendExcludeNodes
pseudonym,pseudonym,...

The nodes list that cannot be used to select the rendezvous point (meeting locations).

SOCKSPort Port

To inform Tor to inform Tor that the connections that are created by programs using SOCKS-protocol need to be blocked through this port. You can set this parameter to zero to not require applications that establish connections according to SOCKS protocol using Tor. (Value as default is 9050).

SOCKSBindAddress
IP[:PORTSOCKSBindAddress IP[:PORT

To establish the linkage to this address to listening to requests for connection from programs that are interacting with the SOCKS protocol (by default, 127.0.0.1). It is also possible to specify a the port (for example, 192.168.0.1:9100), which is obvious, must be "open" through a an appropriate firewall setting on the machine to serve specific reasons. This option may be repeated multiple times in order to perform simultaneously ("parallel") linking to a host with various ports or addresses.

SOCKSPolicy, policy, and policy ,...

It determines the rules for entering an individual server, with an aim of limiting the number of clients' computers that are able to connect to the SOCKS port. Description of these policies are similar to how it works with output rules (see the next section).

TrackHostExits host,.domain ,...

Each value within the list, separated by commas, Tor will track recent connections to hosts that correspond to this particular value. Tor will try to utilize the identical output (locking) node for each one. If a list item of a normal type is preceded with the symbol ". " Then its significance will be interpreted as a reference to the domain as a whole. If one of the list items has only the one "point" and it will display its "universal" connection to all paths. This can prove to be helpful if You often connect to servers which erase all records of Your complete authentication (i.e. make You leave and re-register) while making the TCP/IP-connection modification that is made with one of these servers on Your new IP address following its subsequent modification. Pay attention to the fact that using this option isn't ideal for You because it allows servers to associate history of connection which is made by a particular IP, to Your account. If anyone wants to gather all the data about your time on the server, users who want to achieve this through cookies or any other method that is specifically for the protocol of exchange being utilized.

TrackHostExitsExpire NUMBER

Because servers, as output links in a nodal circuits, are able to begin and stop it at the discretion of the host i.e. in one way or the other - arbitrarilyor randomly It is preferable that any association between the host and the output node is automatically terminated upon the expiration of a certain number of seconds of network activity by the server. By default - 1800 seconds (30 minutes).

So, Tor can be very easily adapted to the current needs.

The current list of Tor instructions is sufficient. The scope of these instructions is beyond the boundaries of the current review. In this review, only the most popular edits and an excerpt of instructions are given. The complete list and details of the directions (in English) you can see on Tor engineering's website.

Visit https://www.torproject.org/tor-manual.html.en

Chapter 9: Use Of Smartphone

If you live located in a nation that blocks websites, such as, China, maybe, you may not be allowed access to certain websites. Tor lets you browse anonymously websites and circumnavigating the censorship of a desktop. Orbot offers Tor for Android and you'll be capable of doing similar things with your smartphones.

If you're connected to the mobile communication of data transmission, or Wi Fi Fi - Orbot is also able to work. Similar to the bundle from the Tor Browser Bundle that is for the personal computer, it's connected to the Tor network Tor and lets you browse anonymously through websites.

If you're dissident in any of these countries like Iran is, this means the government will not be able to locate the person who posted important information on the Internet. It also helps to avoid restriction on the Internet as well as accessing the websites that are particularly useful in countries such like China in China, which has the possibility of restrictions. If you're situated in the USA or anywhere else in the world, this means that your browsing on the Internet are not tied to you and will be kept in databases because of the PRISM or similar software.

The functionality was only available to those who utilized Tor on computer. Now , you can connect to Tor through Android which allows you to use Tor via a mobile phone. Alongside the ability to prevent an intercept by the operator, provider as well as the government also have additional benefits that are offered by Tor for mobile access. For instance, you could utilize Twitter on Android through Tor.

Some dictatorships have restricted access to Twitter and, during demonstrations, you were not accessing information. However, Twitter on Android is able to be set up using Tor. In this case, Twitter will be accessible even if the government blocks access to it.

Connecting to Tor via Orbot

Orbot is the main component of the puzzle. This app on Android is linked to Tor and makes a an internal proxy that other software can access on your smartphone by requesting permission to connect via Tor.

To tweak Orbot is easy. Simply install the program, then open it, and then go through the its configuration master.

If you have access to administrative rights on your Smartphone, Orbot can function as transparent proxy servers. This means that it will make sure that all network traffic go through Tor. If you decide to take this, be aware that certain software can display your real IP-address. If you want to browse anonymously, you should utilize a browser designed to conceal your IPaddresses. If you don't have these rights, then it's okay that you make use of Orbot in conjunction with Orweb and other applications.

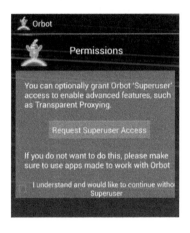

Press the long button on the right side of Orbot And Orbot will join its Tor network. The icon will glow green when the process of connecting to Tor.

Anonymous browsing via Orweb

Once Orbot has been set and started it is possible to use the browser from Orweb for incognito experience. Orweb is well-adapted to working in conjunction with Orbot as well as Tor. For instance, Orweb does not keep records of pages you have visited or other details about websites that you've visited. Orweb also shuts down JavaScript and Flash by default, and also Tor Browser Bundle running on desktops. JavaScript and flash, in theory are able to be utilized by a web-based site to determine of the IP address of your mobile.

To open Orweb directly from Orbot just click on the globe icon in the upper right-hand corner of screen in Orbot. Orweb will be open and show a message that indicates that the user has been connected with Tor If everything is working perfectly. You can now use Orweb's browser Orweb to create an incognito.

Other programs that are compatible with Orbot

Orbot is also able to serve as a proxy-server to other programs. Any application that can support proxies in theory could pass traffic through the proxy servers of Tor Orbot. However, Orbot contains the list of other applications which can be modified to working with him. For instance, you can utilize the reliable communications Gibberbot to search. Tor is an application that is part of DuckDuckGo and to browse the Internet using Firefox for Android as well as with the application Proxy Mobile, or to make the proxies for Twitter to "localhost" or port 8118.

If you are able to access administrative rights and have adjusted transparent proxy servers, then other applications should cooperate with Orbot in theory, however, more secure, if you make use of the software that has been specifically tested for proper use with Tor.

Remember that browsing is significantly slower when you are using Tor in the normal way due to the fact that routing increases the overhead cost. However, if you're looking to browse anonymity to browse the website or browse around the censorship system, this loss in speed is an acceptable price.

Chapter 10: The Reasons Is The Tor Approach A Real Solution?

As we've previously mentioned, TOR is the short name of The Onion Router. It was initially developed by the military was to establish a secure communications network for use by the government. These days the TOR servers permit anyone to be anonymous online. Similar to the onion's peels (the emblem of TOR) the service comes with many different layers (routers) to shift the traffic so that it can disguise your true identity.

Anyone who is looking to discover your identity through TOR will come across an assortment of random TOR servers. This is, in reality, nothing. The reason behind this traffic reversal is because the servers of the TOR network function to conceal who is actually responsible for the activities (you). Therefore, you'll remain hidden from third-party services which track your personal data.

What is the purpose of What is the purpose of TOR?

To give more context for this multi-faceted software tool, here are a few of the scenarios in which TOR might prove useful:

* You're looking for the internet for information, but you're not allowed to do this. Therefore, you must keep your identity private.

* You'd like to avail use of a public computer be sure to keep your data secure. private information leakage.

* You don't want to share any personal data online with ISPs, advertisers websites, or similar sources for collecting data.

* You must avoid the state's censorship or the police (most probably in nations that have these laws) or contribute your knowledge to an organization like WikiLeaks.

How to make TOR work properly

The following guidelines will outline the rules of conduct that will ensure that your experience at TOR is the most enjoyable:

* Install the TOR browser. The installation of the TOR browser running on Windows is

easy. All you have to do is open the TOR Browser Official Download page, search for the most current version and follow the steps to install. I'll go over this procedure in greater detail on Appendix A. For other OS it is necessary to follow the specific steps based on the system you are using. In the appendix that follows, I'll provide some sources for the most commonly encountered scenarios.

* Don't use torrents in conjunction with the TOR. It is important to note that the TOR service is not intended to be used in conjunction with peer-to-peer connections to share files. This means that you are making a poor use the service and slowing other's connections in the process. Additionally, BitTorrent exposes your IP address. Therefore, you'll be sharing your identity through P2P, making TOR useless.

* Do not permit browser plug-ins. Did you create the plug-ins yourself? Most likely, you did not. So, you do not know whether the program will collect information about your identity during its regular use. So,

allowing the programs to run indefinitely can be risky for your privacy.

* Make sure you use HTTPS. The nodes that exit the TOR network is one of the most vulnerable areas for your privacy. TOR secures your data within its network, and hides the origin of your activities. However, your activity that is not part of the network is visible. What do you do? Use a consistent approach to end-to end encryption, like SSL as well as TLS. There is a possibility to utilize HTTPS continuously is a given by simply switching the HTTPS Everywhere add-on, or an equivalent one on an approved website. Websites that are not able to support HTTPS navigation could expose your actions.

Don't open downloaded files using TOR when browsing on the internet. Imagine that you have downloaded a document using the TOR. Then, you innocently click on the name of the document to look at it. What happens? Most likely, your browser will open a compatible document in a different tab. A majority of these documents tend to be shared with Google Drive or similar storage services... that require an account login by the user. This means that you'll be surfing anonymously , and say "Hey I'm John Sanders" at the same at the same time. Is that not clear? Solution: Do not click on things that open further pages using the TOR.

Use bridges. What are you doing in real life if there's a river, and you desperately desire to cross it, but with no contact to the river? You sail a boat! Okay, that's not what I was searching for... the majority times you'll walk across an elevated bridge and use it to travel to the opposite side. The concept is identical when you travel using TOR. You create bridges (relays = safe routes) to eliminate censure. Tor Project Organization Tor Project Organization also provides various bridges to users. These are secure ways to use online without risking your identification. Therefore, you should make consistent use of these.

* Get more users. To be a valued part in the TOR community, don't go online by yourself. Make contact with your friends; inform your friends about TOR. Keep in mind that the more people are connected to TOR the

stronger the network will grow slowly but gradually.

What is the best time to not use What is TOR?

It's not the same thing as being secure online. In the end it is true that the TOR network isn't the ideal solution for every specific scenarios. It is currently in the process of being developed. To utilize TOR you will require a browser that is compatible with this protocol. The protocol isn't damaged yet, but this browser may be the weakest component in this.

Browsers are computer program that connects the user as well as the online network. It is, however, subject to exploits from which someone else might profit from. It was recently reported that the NSA can reveal who is who is behind this TOR network. Naturally, one shouldn't put your money on the chance of escaping authorities using this network.

Thus, illegal use is therefore not "safer" or off the radar when you turn on the TOR. It makes it more difficult for authorities to get to you, and they'll achieve their goal in the end. The TOR network is not a way to make you a criminal. it is not a tool to be used to achieve illegal goals. A misuse of the network can harm the TOR community around the world.

This list contains some non-approved uses for the TOR system:

Large file anonym downloading. This is because you'll most likely utilize P2P services like Torrent. This means that you'll be causing a slowdown to everyone's internet connection and not gaining any benefits, i.e. officials will have the ability keep track of your downloads.

* Attempting to avoid being spied on by the NSA. Why not do this. It's not efficient this way. Just, don't.

* "Securing" your online activities through social networks. The reason is that it's an unlogical choice to utilize services that require ID and try to remain completely anonymous aren't you? In addition, you will not be as secure when you do this; you'll also be vulnerable to service vulnerabilities.

* You shouldn't try to gain access to official websites (governmental and similar). Justification: anyone who uses these services must have an identification.

* Illegal use (children's pornography, unapproved purchases, commerce involving drugs, etc.). The obvious reasons are morally wrong and the possibility of prosecution. Although these are all common due to obvious motives, they are also illegal and could result in you be imprisoned or prosecuted without warning. The environment of TOR makes it simple to use the browser for these types of purposes however these weren't the primary goals of TOR.

As you can see, there are a variety of situations where TOR is the best solution to your issues. In some instances, it's not advised to utilize this type of service in order to avoid to draw the attention of authorities. This is in the event that the NSA as well as the FBI monitor any suspicious activity on the network. A prudent usage of TOR is advised, and is your personal responsibility. Let's look at the way this server network operates in greater depth.

How does TOR work?

A onion has many layers which protect and shape the vegetable. In the same way, TOR is a network composed of multiple routers which function to serve as "layers". Similar to the real counterpart in the real world the outer layer protects the inside, in this instance, your identity. The vegetable that inspires the TOR anonymous service.

Imagine you are required to ship an item that contains important content. What can you do to protect the item during delivery? You can use bubble wrap plastic to wrap the object. After that, you can put on additional layers to guarantee its durability.

TOR is a service that works in a similar method. Your most valuable information is your personal data , and your inquiries through the Internet. To prevent being easily tracked, TOR directs the flow of data through different points within its network. It's the equivalent of having multiple layers of protection to safeguard your identity online. Network nodes are composed of servers and routers all over the globe.

Does it sound easy? You don't require the technical specs for the service. Let's look at the level of security offered by the TOR.

How secure is The Internet of Things?

There is no way to be that is 100% safe in this world. Nothing, not even living beneath the weight of a massive stone can protect you from threats from the outside world. There is therefore no reason to think that TOR is less secure than other options in every situation. While it certainly excels in everyday usage but there have been a few weaknesses revealed over time.

This list provides a summary of the crimes that are involving the TOR network:

The autonomous System (AS) listening. Anyone can spy on the data that flows in or out of TOR network. With a sophisticated tracking system techniques, they can track your identity or even pinpoint where you are. However, this is cannot be done unless you are an expert in hacking techniques.

Exit node listening in. A exit node is an area where TOR is no longer in control over the information that is transmitted to a server i.e. it can connect to any computer around the globe. Thus, an expert can access passwords, e mail accounts, and more from these areas. If you don't use any of these inside TOR, you're protected and secure.

Attack on traffic analysis. While this type of attack isn't able to reveal the user's ID, it may gather data regarding the user within the TOR network. It's not a big issue but.

TOR exit node block. Certain websites prohibit TOR users users from using their services, without having to prove they are.

Therefore, you won't be able edit the Wikipedia when using TOR or to play the BBC on-line player etc.

Bad apple attack. This vulnerability exploits weak services, for instance P2P BitTorrent clients within the TOR. Therefore, don't make use of these networks for "download" anonymously. You are compromising your privacy rather.

Protocols which expose IP addresses. Apart from BitTorrent there are many other protocols that expose the IP address (and consequently your true ID) like P2P tracker communications distributed hash tables etc. These attacks are centered around the vulnerable man in the middle vulnerability.

Sniper attacks. What's more damaging than a denial-of-service (DDoS) threat? A distributed version of DDoS, i.e. an attack on the exit nodes in order that attackers can identify the ones you're using. It works in the following "simple" method by blocking a large sufficient number of nodes, so that anyone using TOR must rely on the one or two that are still operational. What will you gain? An increased chance of identifying the areas "you" use; consequently, a greater chances to discover the person you really are. Of Of course, this isn't exactly a simple method.

Heartbleed bug. This bug was discovered as well as compromised passwords during April of 2014 within the TOR network. The protocols were then removed. But, there is a possibility of identical vulnerabilities again in the near future however, they are not that significant. The primary way to prevent problems like this is to be reliant on hidden services. We will look at a few of them in the book.

Mouse fingerprinting. This is too complicated, but it does work. By comparing the clicks you make on a website using JavaScript to monitor the movements of your mouse, an expert confirmed that it is "feasible" to establish the identity of a person. However, if you don't make use of public computers, this threat is not a concern.

Attack on the circuit fingerprinting. There was some unveiled news of a security flaw that was linked to fingerprints in the TOR network. Therefore, it is recommended that you avoid using bio trackers.

Volume information. The privacy TOR provides you with is not able to hide the volume of data you transfer about. Therefore, if you're being monitored of any type you could be tracked by them movements even through the use of the TOR.

Other. The more complications will be discovered as time passes. Some will be solved immediately; others may take longer, but be remediated in some way. Of course, there's no better than an all-inclusive, never-explored system that protects your identity online. Just like in real daily life.

The list might appear to be as a reason to avoid frequent use of TOR but this isn't the scenario. But, highlighting the potential dangers is my obligation in this intro for the service. The lesson is this: TOR will not guarantee that your online activities are 100% secure. It provides anonymity.

The process is quite simple. To utilize TOR, you will require a browser that is compatible to the service. Visit the TOR Project website to download the latest version of the software tool.

In particular, you could be thinking about making use to the TOR browser. It is most likely the all-arounder choice to begin your journey into privacy online. There are also add-ons to other popular browsers like Mozilla Firefox.

If you're an existing Mozilla user then this add-on might be of interest to you. However, the features aren't as extensive as using the TOR browser, but the ease to install an add-on is not tied to. If you're interested in knowing the advantages and disadvantages of both alternatives, are some tips.

Positives and Negatives of the TOR Browser

The following are the most notable benefits and drawbacks of the use of TOR for navigation.

The positives of TOR

* Most reliable anonymity features. This is a fantastic option to keep your identity concealed for the majority of reasons and

normal use, such as reading stories and not leaving "tracks" in the background.

* Access to Deep Web (or Dark Web). A shady, but potentially dangerous area within the Internet. This "place" isn't searched by search engines, consequently, it's still in the dark. Furthermore it is a place where illegal activities are carried out within this online "black space". Be very cautious regarding the things you are doing and the websites you go when you are on this website - else you could be in problems with the authorities.

* The majority of novice "hackers" aren't able to recognize your vulnerability. It instantly gives you an unbeatable protection against those who have watched numerous hacking shows or films. They'll never attack you, however a genuine hacker might. Don't think you're completely safe.

* More. TOR establishes a standard for private browsing. It is portable, and access is concealed through .onion websites.

The downsides of TOR

* Performance isn't for people who are picky. While the performance has improved over the last few years, it's not exactly like the normal way of browsing. But, there has to be some cost to pay for this extra security layer of anonymity, don't consider?

* You aren't completely safe or unnoticed. Authorities are monitoring you. Don't forget that. NSA and FBI are taking this practice extremely seriously. If you get caught on the wrong websites, or even accidentally on the Dark Web; they will locate that you. Beware of them at all costs. Be careful, but don't be worried even if you make an error (i.e. accessing a prohibited website accidentally) and you take the correct action (A. immediately leave or B. Inform the website) There is no problem for you. There's no reason to conceal anything, either. Of course, they'll be able to ask questions. Paraphrasing ol' uncle Ben "great power demands great responsibility". Therefore, it's always your decision to act in the right way as citizens.

* Reputation. The previous problems with authorities as a result of the misuse of TOR

led to the present. Unfortunately the TOR community is a burden in its own back. But, it's an incredibly powerful tool; it has gained a lot of popularity with activists and journalists around the world. There's always room for useful contributions from the use of TOR.

The network is very low latency. This is a very common problem in the TOR network. You must be patient. The patient will discover what he is looking for.

Mozilla Plus-on positives and negatives

These are the most impressive advantages of using this Mozilla extensions of TOR service. Let's look at these features so that you can decide on your own if it's worth the effort for you.

The positives of Mozilla Add-on

* Open source community that is free. All Mozilla products are covered by a public license. Therefore, any developer can help in the development quality of service. Additionally, anyone can make certain that it delivers what it promises. Transparency is

an essential factor in open source communities.

* Customization. As is the norm in Mozilla it is possible to personalize the add-on to your personal preferences. Thus, you'll enjoy an entirely unique experience when browsing.

• Community Support. Many users prefer the TOR variant of the program, especially Mozilla fans. Mozilla Firefox is an all-rounder well-built browser that is compatible with the majority of platforms and devices. If you decide to upgrade your device later you won't have to alter the experience in any way. In addition, the community support is always a benefit.

* More. There's a wide array of things to note for example, like a robust continuous development, strong HTML5 support as well as syncs to devices, tagging bookmarks, the ability to read without distractions, which can enhance the scanning experience, automatic updates, quick bookmark management that uses the least memory and CPU load , and integration with Pocket in order to store pages while on the move.

Some disadvantages to Mozilla Add-on

* Installing extensions demands restart. This is a problem in the event that not all extensions supports non-restart running. Therefore, it is important to be patient with it.

Performance is slow in certain OSs. Some users have noted that performance in OS X is much slower than Windows as well as Linux. Be cautious when you're an Mac user.

The longevity of extensions is at stake. Extensions' life span is not a secret within Firefox. The reason is Firefox operates at its own style and pace, while the extensions follow their own. Therefore, it is possible that an extension gets abandoned and then becomes ineffective as it's no longer in use. Therefore, if you are relying on an extension to enhance your experience, you should also take into consideration the work of the developers during the update.

* More. There are other disadvantages of using Firefox to TOR navigation, but they're usually specific to the situation (i.e. it only impacts a small percentage of users in

certain situations). For instance, there's enterprise support. This is not necessary when you're an individual accessing the internet from your home. Furthermore, there's no HDiDPI support, and on those high-resolution screens, the icons appear blurry. If you're not browsing on the 4K resolution and you don't see it, you will not notice.

How to Begin

You now have either a TOR-based browser or the add-on version available or possibly both! It's time to test the basics of online activities to discover the differences between browsing with no anonymous features.

In the next section, we'll provide an overview of the most appropriate candidate for TOR services. Take a look at that section to learn what you should anticipate from this service, and what you shouldn't be expecting. Be sure to make responsible use of this fantastic tool.

Are you a suitable user for Thermostats?

We've already said that TOR isn't suitable for all available. I think the most effective way to comprehend the intended audience of TOR is to review the most advised practices that are compatible with the service. Let's take a look at the most effective practices to be an efficient user of TOR.

This chapter examines some of the important methods of security and privacy that are appropriate for internet users.

Modify Your Operative System (OS)

"Winbugs" vulnerabilities are famous. The most used OS around the globe was developed to be user-friendly. Therefore, there are vulnerabilities that haven't been averted. Switching to a different OS might sound like a radical idea, but over the long run you will realize that the advantages are substantial. Linux systems are fully compatible with the TOR.

In reality there are several distributions that can be configured to work with TOR like Tails as well as Whonix. Mac has also been adapted to TOR It is all you need to do is for the web browser to be installed, and you're good to go and it's as secure as Windows.

Be updated

Whatever the OS you're using make sure that you keep your system up-to-date. This is a crucial aspect that must not be neglected given that older versions have vulnerabilities well-known by malicious hackers. Therefore, make sure that everything is as current as you can. This includes your OS as well as the TOR browser.

HTTPS Everywhere

It's possible that you are wondering how to utilize https on the case of a site that only has an the http version. There is an option called HTTPS Everywhere. The websites that are supported automatically switch to HTTPS-mode browsing by the help of this extension.

Secure Your Data

So, why should you bother being anonymous online if you can not gain from additional security features such as an onion - does that sound familiar? LUKS along with TrueCrypt are two programs that permit encryption in Linux systems. The more encryptionyou can use, the better security.

Be careful with the Tor Bundle

Contrary to what many believe contrary to popular belief, this bundle could be harmful. Yes, it is an additional layer, however, there are weaknesses which have been identified through the FBI. Be wary of using too many of these tools.

Disable Java Script, Flash & Java

These malicious scripts could give your personal details to others without giving you a opportunity to prevent it. Therefore, turn them off and enjoy your surfing experience in complete privacy.

There is no P2P

It's probably the third time I have mentioned this, yet it's essential to know. If you are using peer-to-peer networks, you are exposed to.

No Cookies and Local Data

Add-ons, such as Self-Destructing Cookies will automatically delete those unwanted pieces of information. They can store personal information which means you'll remain anonymous with none of them.

No real email

Don't ever create us of actual accounts of your own on the TOR. If you do, you'll end up telling people "Hey Hello World. I'm anonymous. Signatureof John Smith." What's the issue?

No Google

The most used search engine on the planet actively collects personal information. Therefore, you'll be more secure using a different engine, such as DuckDuckGo and StartPage.

Do the Legal Stuff

The last thing to do is adhere to the laws and regulations when surfing the internet. There are two agencies that monitor your online activity. NSA as well as the FBI constantly monitoring to determine who's making a wrong use of any website that is similar to this. It is best not to draw unneeded attention to this issue by adhering to all guidelines.

If you're not complying with the rules of your government, make sure to keep your activities as private as you can. You can only use obscure services to create secure communication channels or to send files without revealing your identity or otherwise. One of the most secure methods to use TOR is to browse the news without interfacing with any service, website or other. If you choose this route, bear in your mind how much interaction you'll have with websites that you use, the more stealthier you be.

In the next part, we'll look at the types of information that you can access when browsing through the TOR. Look closely at the plethora of possibilities this network offers you in your hands; beginning at an introduction to the Hidden Wiki.

.

Hidden Wiki

The website is similar in layout as Wikipedia but it's the TOR equivalent. It allows you to browse the various categories that you're interested in to discover the choices that you may have. It is the Hidden Wiki is that all .onion sites of the Deep Web are listed, covering a variety of subjects. The search for The Hidden Wiki is simple, simply type it in DuckDuckGo Search Engine using TOR and you're just one mouse click away from accessing the Wiki-hub.

There is a extensive list of websites on this site including:

1. Introduction Points. These are the primary websites that users will see at the beginning of their journey when they open using the TOR browser. In most cases you will arrive on these sites using StarPage the built-in search engine that works with TOR.

* News. This is among the major reasons why you should make use of the TOR. You

can avoid government prohibitions, restrictions, and bans to read world news with no restrictions. There are portals within TOR to stay up-to-date with the most recent news. Check the ones that are regularly updated.

* History. Contrary to popular belief, History is not taught in the same manner across all nation. Furthermore, there are countries that alter historical events in accordance with the political beliefs of their citizens, personal beliefs and so on. So, as it is with newsreports, TOR provides an objective view of the truth.

* Business Services. Are you looking to buy something without having to reveal your identity? Of course, there could be more convenient ways. But, the growth of Bitcoins allows you to shop as anonymously as is possible within the online world. Naturally, you can be tracked by someone who tracks your actions, but it'll be much more difficult to track your steps within TOR on a daily basis.

* Forums. Are you looking to discuss a the subject in a private manner? There are

applications that provide this option, like the Q&A community which Quora currently offers in English as well as Spanish. On Quora, you can submit your question anonymously and respond to the community. The people who run the site know what you're talking about, obviously. If you'd like to go a step further in protecting your personal information You could consider using forums in TOR and other private services.

• Other "dangerous" subjects like Hack, Phreak, Anarchy, Warez, Virus, and Crack among others. Avoid any illegal site while browsing within the TOR. In certain instances, the websites could appear without being able to block it. But, being aware of the content is the best method to avoid these sources of illegal activity as much as is possible. It's possible that you won't be able to block pop-ups completely, but you will be able to stop them immediately without having to interact in any way. Similar principles apply to regular usage of Internet.

The primary directory of websites within TOR is called the Hidden Wiki. Although the lists of websites on the Hidden Wiki may not always be kept up to date, as some may have been offline for a while however, it's still a good source for onion addresses. Be aware that the reliability of sites within the TOR network is less stable than the other internet. Internet.

When you discover websites that you like I would suggest that you save them in a safe place. As opposed to regular web browsing TOR URLs may be tough to remember. Furthermore, directories can change dramatically at any time, forcing you to look for a long time. Bookmarks that are used in a smart way is always an excellent source of address information for you.

Onion Chat

There are chat rooms that allow anonymous chat which is called OnionChat. They are provided through the TOR service to ensure that the the service lasts longer. If you happen to know anyone who is using the TOR service, you can use these chats to be a way of communicating.

In the end, why bother with anonymity when you're not going completely commit to it, right? The use of nicknames is a regular practice when chatting on TOR. Never use real life names.

New Yorker Strongbox

This website is for secure messages that writers can use to communicate messages or files to the editorial staff of The New Yorker. It's completely private. You will receive an account number when you sign in to make sure that you don't divulge personal information from your part. Many bloggers who run websites on the TOR network do not post, which means you might not receive new feeds from bloggers you are interested in reading in this area of the world wide internet.

Other pertinent mentions

However, TOR is not a tool that everybody can master currently. Once you're in the Deep Web, you are completely at your own risk. Therefore, the feeling of being lonely is greater than when you surf normal. Furthermore, the extremely small life span of websites is not enough in making this feeling less infamous.

Secure Browsing

Remember that TOR can only be as efficient as your browsing habits relate to your the privacy of your ID. By default, TOR does not communicate with Google. The most well-known search engine in the world is a sytem for private information of its users; it keeps vast logs of all the things you use the service to search for. However, TOR connects to "Start Page" which acts as a intermediary that connects you to Google and ensures that the session is not traceable. Through using Start Page as a relay you can ensure that your searches are not linked and traceable to the source of your activity, i.e. you. It's not through any easy means. I don't take into account the NSA level of inspection to make this particular observation.

TOR is not able to regulate the actions of extensions scripts, websites and scripts. Thus, the best way to minimize the risk is to avoid any happening. As a default setting, TOR browser doesn't permit such activities.

It is not advised to alter this setting at the middle of your journey to prevent unnecessary exposure.

Are you aware that you can stream without registering? While it might not always work, Youtube is now offering an HTML5 beta service in operation. The standard flash videos on the famous site aren't available when you use TOR due to security issues within Flash protocols. Therefore, you won't have the same experience well as you would in normal browsing mode.

Not to mention, TOR will warn you about files and documents which could reveal details regarding your ID in a casual manner or deliberately. It is advised to read an in-depth look at these warnings. You'll find that it's not advisable to risk anything.

Anonymous Messaging

When you are in normal mode of navigation, there's nothing quite like private messages. Any messaging app you can consider has logs of monitoring for the "private" chats, for example, Google chats, Facebook, Skype, and the like. How can we get away from "Big Big Brother" for a second to communicate with someone who is not disclosed?

It's a good thing that TorChat lets you do just this. You can use this anonym chat application as an extension. To use this service, visit the official site of to download and install the file. You will be able to download an executable files that runs the program.

The chat client functions the same like any other messaging application and you'll appreciate the user-friendly layout. The main difference between this and other messaging programs is that you'll be identified with a random set of characters, instead of your actual name displayed on the application. You can change the names of your contacts in order to help you establish conversations. Codenames are highly recommended. Additionally, since the service operates within the TOR background, no one within this network is able identify who you are communicating with at any time.

Anonymous E-mail

So, what happens to emails? It is fine to send messages that are not identified however, there are times when you might have to send an email instead. In the list of "hidden services" that are available within TOR it is of course, email. But, be aware that any hidden service is only accessible within TOR. it could not be part of within the network without the risk of compromising identification information.

The Deep Web

We will talk more concerning hidden features in this article. Additionally, a second review of security issues and limitations will be conducted.

Hidden Services

We've already discussed the most popular hidden services that users would like to think about when using the TOR. When you read this article you will notice that more of these services are being created and others are being removed and the Deep Web is in constant evolution.

Why use hidden services?

Utilizing a hidden feature in TOR is an ideal solution for privacy issues since it's never subject to outside surveillance (from the normal Internet). Therefore, these services aren't shut down or disabled. Regular Internet services are quickly refused.

Let's examine this using an illustration. I could transmit massive data packets to the router of my neighbor in order to block access to Internet because of saturation. That is I could interfere with the connection with the router as well as the computer my neighbor uses. But I'm not, and I'm a good neighbor!

But, if the neighbor in question wanted their service private, and connect two computers to their house and a wire is sufficient to block any intrusion from outside the house. Turn off your Wi-Fi. Do you see where I am going?

Utilizing bookmarks

Making bookmarks consistent is an essential skill when browsing the Deep Web. Be aware that a typical hyperlink in TOR network appears like this:

http ://[bunch of random characters here].onion

Also, it is difficult to recall websites or service addresses within TOR because of the random characters that appear in URLs. Therefore, bookmarks are your most reliable companions to navigate the way you wish to.

There isn't Google on the Deep Web nor are the pages indexed by standard search engines. So, how do you get information from the Deep Web?

TorSearch

This is the equivalent of Google within TOR however, it does not constantly glean personal data from you for every search. This service is based on the same concept that the Google search engine does.

After the addition of TorSearch the volume of traffic was rapidly increasing in the network. Today, it has grown significantly from the beginning. It's now easier than ever before to locate the hidden service, site, or generally anything else you'd like discover on the Deep Web.

The hidden Wiki is still a resource for all who wish to explore the categories to discover their own interests. This site is controlled to only display public and "acceptable" content that is acceptable to every public.

Other concerns to take into consideration

While TOR is a wonderful solution to protect your privacy online however, it's not completely free of weaknesses. If you use the TOR browser is an altered version of the Mozilla Firefox counterpart, the similar vulnerabilities could be used. Furthermore, you may be vulnerable to Trojans even when you use the browser as part of normal browsing. In the end, you don't think about when or the place you might encounter an infection, do you?

Another disadvantage of using the TOR browser is that you'll be at an attention from the NSA as well as any other similar governmental agency. They will be able to monitor the activities of your browser and determine your internet activities. So, you need be aware of the websites you visit. If you are a visitor on illicit website can cause problems with authorities. Beware, you've been warned.

Websites to be aware of

In addition to other than the Hidden Wiki, you also have search engines for the products within the TOR. But, be aware of frauds of all kinds. It's a good idea to stay out of some of these search engines to avoid problems with authorities, like Grams which is a well-known website that has gained traction within TOR. The search engine is an exact replica of Google that is linked to the crypto-currency market for drugs. So, be cautious for yourself if you are trying to find an item using that tool.

Not least, take a look at the references of a product before making a decision. The community of users is very active on certain websites. However it is a good idea to be able to get enough information about the service or website, do not make use of their offerings.

Security concerns are also raised about the content

Accessing content on TOR is more risky than surfing on the normal web. We conduct an assessment of security of the type of content you might find on the network.

* Content that is dynamic. These kinds of websites require questions or forms to browse. You'll need prior experience with domains in order to browse these sites.

* Content that is unlinked. These sites are not indexable through search engines. Additionally, there aren't backlinks or internal links on these websites.

* Private websites. Websites that require a login in order to access. Be aware of possible vulnerability to private data and fake passwords which can let you disclose your personal details about yourself to.

Contextual websites. The access context for these website may differ. For example, they could depend on clients who have previously visited.

* Content with limited access. These websites use standards for robot exclusion, CAPTCHAs, and the similar. Therefore, they

are not indexed by the search engine on a regular basis.

* Content that is scripted. These kinds of websites require hyperlinks created using JavaScript or other software applications like Flash, Ajax, and related.

* Software. Applications or programs are needed to access these types of sites.

• Web archive. Certain archives let you look up an older versions of websites. They keep older versions of the site to be able to check them again in the future.

As you will see, the pages appears in various formats on the internet, either through regular browsing or in the TOR. Therefore, being able to distinguish between the two is essential to know how secure website or service could be. As a rule do not install executables or software that you discover on the internet so that you are able to limit the risks.

the Future of the Deep Web

It is unfortunate that it has to be the current political turmoil that has been affecting the Middle East one of the major reasons behind the increased use of the TOR network. Over the last couple of years, the possibilities of using of this software and network has been demonstrated frequently. The users have been able to escape the government's censorship due to many reasons, including simply to watch cat videos on Facebook.

But there's much more to the Deep Web. It is impossible to estimate the amount of data stored within the Deep Web; the section of the Internet that isn't indexed through search engines. A few years ago, TOR was estimated to contain just some Petabytes of data. But, the number could have increased dramatically over the past few years. It could be ten times or a hundred times or even greater than what was initially estimated.

You will be able to see by reading the last chapter of the appendix to this book, how simple it is to build your own server using the TOR program. This is not rocket science anymore; it's a simple procedure to follow right now. There are a lot of tutorials that you can use to create a secret service that you can create your own. As you're capable of doing it, so can people from all across the globe. This is the reason why the content within TOR is increasing exponentially. Users enjoy the advantages of anonymity online, and feel a of a feeling of security.

Don't forget that TOR doesn't really safeguard you in the same way as Firefox does. Firefox browser. Similar vulnerabilities could have a bearing on both in the near future. Furthermore, since the popularity is in the high-end of the usage of TOR and other hacker-friendly services, more hackers could attempt to obtain personal information on the network through illicit purposes. Therefore, it's something to be cautious about. It is a place to be cautious about. Deep Web is not as safe, delicate, and content-wise like the normal counterpart. It's similar to it's real, but it's not subject to any guidelines or supervision.

This is a problem If you realize what "simple" you can find fraudsters online. There are a myriad of websites that attempt to convince you to make poor decisions on your spending money. This can end with you spending money for nothing in most instances. This can undermine the credibility of trustworthy sellers. But, that's how the Internet is, and it is not likely to change anytime soon.

Does that mean you can not utilize the WWW? Absolutely not. There is also a chance of being injured if you leave your home, or even stay at home. If you apply the same bias reasoning, it is best to remain in bed to stay "safer". I'm online almost every day. I've been doing this for more than a decade. I'll be doing the same within a decade. What do you think? I can't wait to watch what an incredible innovations in human history develops over the coming years; the Internet has changed the way we live our lives totally different.

Imagine how communication was half an century ago, and then think about the way they're now. Remember when purchasing was not possible prior to the Internet and you were required to go to the store at every store until you could find the perfect product you wanted. Imagine driving time ago, without GPS only yourself and your road perhaps the map.

Privacy Beserches

Nowadays, we get stories from around the world in a flash live stream, keep live communication and videoconferences, among other things. In this environment it's not impossible to want some privacy, isn't it? That's why TOR was developed.

A large number of users have used TOR to be activists or part of the anonymous opposition faction's forces. Remember that in certain countries, a violation of certain laws governing censorship could lead to punishment and perhaps even prison. Thus the use of TOR is a must for those who don't wish to be oppressed.

Chapter 11: Deep Web And Dark Web What Are They?

Before talking about the definition of what is a TOR browser is and how you can utilize it and reach the depths of the Internet using it, we must be aware of the meaning of Deep Web and Dark Web is. The two terms are frequently misunderstood.

What exactly do you think Deep Web?

In essence, this is an additional layer of the Internet which we cannot access using a standard browser or require the proper permissions for accessing it.

The information this particular corner of the Internet is hidden are the government databases, medical reports and information on the users of different websites, cloud-based content and private videos on YouTube and more.

The majority of statistics indicate that an average Internet user can access just 4 percent of the data that is available in the Internet. The whole network is frequently described as a huge glacier, with only a small portion of it extends over the ocean's surface. In this scenario the ocean is called part of the Deep Web.

What exactly is Dark Web? Dark Web, is a small portion that is part of Deep Web, size comparable to a web's surface. Its Dark Web pages are designed to let users browse them. However, tools, such as the TOR browser are required to do this. The majority of information we have about this type of layer is reports about its criminality.

It isn't true there is a misconception that the Dark Web is only used by criminals. Many of the people that we meet here are simply concerned about their privacy and desire to conceal their identity from the world.

Also, we should not forget that governments in certain nations are oppressive to their citizens and restrict their freedom of speech through the Internet. In this situation, the Dark Web may be the only method to avoid restrictions on speech.

Within the Dark Web we can find forums that connect many Human Rights Defenders and human rights activists of countries where human rights are not respected.

Crime and Dark Web

Its TOR browser, essential to access this portion of the Internet provides us with anonymity and complete protection for our personal privacy. This provides the perfect conditions for the spread of crime. There are numerous websites selling illegal products ranging from weapons, drugs or stolen credit cards, to fake identities, passports and licenses. Although some fencees will fulfill the terms of the contract and provide us with the goods ordered, most of them are owned by criminals. The most preferred method of payment on the Dark Web is BitCoin, which provides the criminal with greater certainty that we will never discover who he is or where he's from.

BitCoin transactions are completely private and the connection between customer and vendor is effectively secured. Add on top of that the fact that we're trying to purchase illicit goods and it becomes more difficult to deceive us.

We will not report to the police, and we don't mention -"Hi, Mr. Politician, I tried to buy myself a bag of cannabis as well as an unauthentic passport, however I was scammed. Even if we do purchase something that is legal but we cannot be sure that the seller is honest. The most effective advice I can offer you is to stay clear of all transactions that occur using the Dark Web.

The browser TOR

The Onion Router (The Onion Router) is a completely free open source browser designed for anonyme and uncensored communications. The name TOR comes in the onion routing that was developed in the hands of two US Naval Research Laboratory employees. The aim was to secure the communication of American federal agencies. The onion routing technology was purchased by DARPA (Defense Advanced Research Project Agency) in 1997, who implemented it to be used by military personnel of the US Army.

The initial version of the web browser that were based on onion routing were launched in 2002. In 2004 2004 the TOR code was made available under a no-cost licence to

Roger Dingledine and Nick Mathewson who continued to work on this TOR project.

Does TOR provide complete anonymity?

Yes, provided that we take every precaution not to disclose our identity, or what device we're using. Any information disclosed by TOR users is due to their negligence or a lack of awareness.

ignorance. All you need to download the file through to the Dark Web through a different browser other than TOR, and our true location is no longer secret.

Security settings

The TOR browser comes with three levels of protection to users three levels of protection: standard, safer, and the safest.

1. Standard All functions of the browser as well as websites are activated, which assures the full functionality of websites. Only use the standard settings when visiting trusted and known websites.

2. Safer - removes potentially risky functions on websites. Certain symbols or fonts may not work correctly and multimedia, such as

videos and audio only begin with a click. Additionally, Java scripts are not available on all pages that are not HTTPS.

3. The most secure option allows websites to utilize only their most basic functions. All scripts on pages even those that use HTTPS protocol, do not begin. A lot of pages won't function when this is the case.

Onion routing

The name comes from the method by which data is transferred via in the TOR network. On the onion network, messages are transmitted using encryption layers, which are similar to onion layers. The data transmitted by the user will not directly reach the recipient. The package has to travel through various randomly selected computers in the network. The data is encrypted at first before being transmitted to the network's nodes, also known as onion routers. The person who sends the information in this type of call remains anonymous because each intermediary is only aware of the location of the prior and next nodes. Further details are accessible via the official TOR project's website.

https://www. torproject. org/

Tails is an operating system that was created to ensure privacy

Tails is a no-cost, open-source operating system that is real-time and that is focused on protecting their privacy. What exactly is a live system? It's the name of an operating system which operates directly from a flash drive , or CD/DVD. It is then loaded into the RAM of your computer.

The system in question does not make use of a hard disk, and thus isn't connected to the data on our computers. When you switch off the system it will erase the temporary files, passwords saved cookies, and other data that leaves a footprint on us is erased.

In the scenario of Tails in the case of Tails, the remaining space on the pednrive gets used to store disk space. This lets us save files, like images or presentations created using the program included. Tails lets us select the files or settings of our program that are crucial to us. It it saves them in a secure space, meaning we do not have to be concerned about losing these files. Other data is erased when the system is turned off, so that there are no traces of data are left on the ground.

Is it really Tails?

Tails is a basic and highly functioning safe operating system. It includes an integrated TOR browser. It also has Thunderbird emails that secure messages, LibreOffice, graphics programs like GIMP and InkScape and the OnionShare application that can share folders and files through the TOR network. You can also get all the applications including calendars, calculators notespad, calculator and many more.

The built-in browser has various plugins which can be activated at any moment and easily set up.

NoScript can be described as an extension which, when enabled, blocks any scripts that are launched whenever the webpage is open. They could be ads, animations, sound effects, that link to external websites or links that are infected by malware, or tracking files. However, many websites cannot function properly without the scripts disabled therefore, if we are confident about an online site, we may gradually increase the privileges it has until we can access it with ease.

UBlock is a completely free open-source extension which blocks any advertisements (including ones that have been affected) and provides us with the option to filter and block any content we want to.

HTTPS Everywhere, a plugin which forces websites to work in a safer encrypted HTTPS

connection. If a website doesn't support this protocol, it might be blocked.

Conclusion

You are now equipped with the information that you will need to be able to enter the realm of the Dark Web. Here are the steps to make this process as secure as you can and be a responsible and safe user that doesn't put your device or yourself in danger. Additionally, here are few other important points to keep in mind that are listed below.

1.Providing you're making use of Tor, which is the Tor browser, then you may actually be more secure when surfing the Dark Web than your normal online activities. It is configured to provide security against privacy threats that aren't addressed by regular browsers.

2.If you do register on a site don't use your real e-mail address, your real name or username. Create a fake identity and, whatever you do, stay away from using a credit card. you're not able to claim and may be required to do some awkward conversations to answer when the charges show up.

3.If you're concerned that your activities online could be a red flag to a higher authority, then be calm, there's such a lot of online activity Dark Web that, unless you are in an strict country it is highly unlikely to flag and attract attention. If this is an issue, you may want to connect to an VPN prior to making a connection to Tor.

4.If you are absolutely required to download something, make sure you don't download anything unless you truly have to, then ensure that you are protected with a reputable antivirus software like VirusTotal. Everything you download could cause harm to your device. It is essential to be sure that you'll be secure from virus. If you see an alert, go around. Don't continue with the download.

5.And lastly, apply the common sense in all that you undertake. In any activity you take on, keep in mind that when something seems too promising in fact, it is. Do not expose yourself as well as your equipment. If you see a random person being extremely friend-like, is he your ideal friend? Most likely not, but keep in mind your basic sense and instincts. It can be very beneficial when used properly and will offer greater protection better than any antivirus or other security software (but it is true that you need these tools for protection).

Keep in mind that once have mastered Tor as well as its hidden features, you are well-equipped to use the internet in a day-to-day basis. Therefore, you should develop your abilities and utilize the services effectively.

The ability to browse the Internet without revealing your identity is becoming of growing importance. It gives you the capability to accomplish tasks you'd normally not be able to do and gives an extra boost of confidence that you didn't realize you needed when it comes to your research abilities.

What is the significance of this and what to do is the focus in this guide.

The book is comprehensively covered how this is done and it can be done successfully.

It has shown how you can remain anonymous when using this technology that plays such a major roles of our daily lives.

It's difficult to remain completely anonymous in the present time and Tor Browser assists in this.

It is constantly advising you of the steps you must take before you go on the Dark Web. This Dark Web is a dangerous location, and if don't know what to do to enter it with care, you could end up in danger. Once you've learned about the risks you could confront, you are able to prepare against them.

Happy Surfing!